Adam Leroy Jones

LOGIC

INDUCTIVE AND DEDUCTIVE

AN INTRODUCTION TO SCIENTIFIC METHOD

Elibron Classics
www.elibron.com

Elibron Classics series.

ISBN 1-4021-7130-7 (paperback)
ISBN 1-4021-1322-6 (hardcover)

This Elibron Classics Replica Edition is an unabridged facsimile of the edition published in 1909 by Henry Holt and Company, New York.

LOGIC

INDUCTIVE AND DEDUCTIVE

AN INTRODUCTION TO SCIENTIFIC METHOD

BY

ADAM LEROY JONES, Ph. D.

ADJUNCT PROFESSOR OF PHILOSOPHY
IN COLUMBIA UNIVERSITY

NEW YORK
HENRY HOLT AND COMPANY

5

To

L. S. M.

PREFACE

THIS book is intended as a text-book and not at all as a contribution to logical theory. It aims to present an outline of scientific method as briefly and as concretely as possible. It is not designed to serve as an introduction to general philosophy. Its chief claim to novelty is in the arrangement of the subject matter. The traditional arrangement in which the deductive processes are presented first usually leaves with the student the impression that method is chiefly deduction, and that there is no very close connection between this and the rest of subject. The arrangement, which is here adopted, was selected on pedagogical grounds and not in the interests of any epistemological theory.

The justification for dogmatic statements on disputed points is also pedagogical. Argument on such points in a text-book usually fails to interest the student and often tends to make him think that the whole subject is in an uncertain state and mostly a matter of opinion. Some subjects are treated much more briefly than they deserve, but I wished to keep them in due proportion with the rest.

Fallacies are first discussed along with the processes with which they are connected, but they are all brought together in a later chapter. Many of the exercises are new, but I have also drawn freely from other text-books. The longer exercises at the end of the book give the student an opportunity to bring to bear almost the whole of scientific method, and for this reason they seem to me to be very important.

My indebtedness to Jevons, Hyslop, Mill, and Bowley will be obvious. I owe much to Aikins' Principles of Logic; his broader treatment of many topics and his chapters on Testimony, Averages, Statistics, etc., were very suggestive. Sidgwick's The Use of Words in Reasoning, Creighton's treatment of the Figures of the Syllogism in his Introduction to Logic, Hibben's use of the idea of system in his Logic and Cramer's The Method of Darwin, were also suggestive. I have tried to give credit in each case in which I am conscious of having borrowed.

I am much indebted to three of my former colleagues in Princeton University: to Professor W. T. Marvin for going over the whole of the copy and giving me much useful advice, and to Professors W. H. Sheldon and E. M. Rankin for assistance with the proof; and to my colleagues, Professors Woodbridge and Montague, for many valuable discussions of logical problems.

A. L. J.

New York, *April*, 1909.

CONTENTS

PART I

AN OUTLINE OF SCIENTIFIC METHOD

PART I

AN OUTLINE OF SCIENTIFIC METHOD

"So the vast results obtained by science are won . . . by no mental processes other than those which are practised in every one of the humblest and meanest affairs of life. A detective policeman discovers a burglar from the marks made by his shoe, by a mental process identical with that by which Cuvier restored the extinct animals of Montmartre from the fragments of their bones. Nor does the process of induction and deduction by which a lady, finding a stain of a peculiar color upon her dress, concludes that somebody has upset the inkstand thereon, differ in any way, in kind, from that by which Adams and Leverrier discovered a new planet. The man of science, in fact, simply uses with scrupulous exactness the methods which we all habitually and at every moment use carelessly; and the man of business must as much avail himself of the scientific method—must be as truly a man of science—as the veriest bookworm of us all." [1]

It is of course true that the conclusions of science are often in disagreement with those of common sense, but the disagreement is due to the difference in the thoroughness and completeness with which the facts have been examined. In many cases the common sense of to-day is simply the science of yesterday, for common sense is usually very conservative, and often regards the novelty of a conclusion as an argument against it.

Induction and Deduction included in Scientific Method.—Scientific method being simply a more thorough application of principles universally employed in

[1] Huxley, *The Educational Value of the Natural History Sciences.*

reasoning, a good means of getting a general view of those principles will be to examine the procedure of science. It includes both formal and inductive logic. Formal or deductive logic is simply one part of scientific method; hence any exposition of scientific method will include an examination of deduction. Sometimes induction is identified with scientific method, but it is often used in a narrower sense; and, in any case, it might seem to exclude deduction, which is an essential part of complete scientific method: therefore it is less confusing to think of induction as simply a part of scientific method. Inductive and deductive reasoning are constituent elements in a single system. For purposes of study, it will be advisable to break up the system into several parts. The first of these parts will include the processes and principles involved in acquiring a knowledge of facts; the second, those employed in the classification of facts; the third, includes the discovery and formulation of laws; and the fourth, the testing of these laws and their further organization and application. Each of these processes will be found to involve a number of subsidiary processes. As they are parts of a system, they are, of course, mutually dependent; each leads up to or implies the others.

The Beginning of Knowledge.—Nowadays there is almost universal agreement to the statement that all knowledge begins in the perception of concrete facts. It has sometimes been thought that the mind began its career with a capital stock of knowledge in the form of "innate" ideas or principles. But no one now maintains that there is any knowledge before experience begins. That, however, is no warrant for the conclusion which John Locke drew.[2] He held that the mind, at the beginning of its history, is like a sheet of white paper or a waxen tablet or an empty cabinet, and that experience, like some external force, writes upon the tablet or fills the cabinet. To him the mind seemed to be passive in the acquisition of knowledge, able at most to combine and analyze its sensations and ideas. Immanuel Kant, on the other hand, contended[3] that, even in those mental operations in which the mind is seemingly least active, it is contributing essential elements; that it makes knowledge, as it were, out of the material which is furnished from without; it cannot operate without material, hence there is no knowledge before sense-experience begins; but this sense-experience itself is, in his view, a product of the mind's activity. We cannot pursue this question any further; our concern is not with the philosophical problem of the ultimate source of knowledge; it is enough for our purposes to know that knowledge begins in concrete experience, in perception, in knowing sounds and colors, odors, moving objects, pains, pleasures, emotions, and so on.

Natural Sciences and Others.—Things and events and relations in the external world constitute the data of what are sometimes known as the "natural" sciences, such as biology, physics and chemistry. Mental facts and their relations make up the data of psychology; they are quite as concrete in their way as any physical facts, and the methods employed by psychol-

[2] In his *Essay on the Human Understanding*.
[3] In his *Critique of the Pure Reason*.

ogists are the same as those used in the physical sciences.

Such sciences illustrate almost all the processes employed in the acquisition of knowledge, whereas a science like mathematics makes most use of a few of them, which it applies and elaborates with great thoroughness. We shall attempt to follow, as closely as may be, the stages in building up knowledge as they appear in the natural sciences. As in knowledge generally, these sciences begin with the perception of facts, external or internal. Then sooner or later they proceed to classify and organize the knowledge thus gained.

The Sources of Knowledge, Direct and Indirect.—Perception of concrete facts comes first as a source of knowledge, or rather as the primitive form of knowledge. Its limitations are obvious; it is often far from clear; it is frequently mistaken; it embraces comparatively few facts at any one time, and it does not extend beyond the present, or, at most, the immediate past. If we had to depend upon it alone, we could never get together a body of knowledge. It is possible dimly to picture a mind which could be aware only of what was immediately present in time and space; its knowledge would be rudimentary, and without knowledge of something besides the present, the present itself would be meaningless. In all but the lowest types of consciousness there is a constant use of *indirect means to knowledge*. *Memory* is the first of these.[4] Memory restores a larger or smaller part of the

[4] It might perhaps be said that memory is direct knowledge of the past, and this is true in a sense; but the dependence of memory upon previous perception, the fact that we do not remember what we have not previously perceived, shows that it is also indirect.

knowledge previously gained in perception, and thus makes it possible to draw upon past as well as present experience.

Another indirect means to knowledge is the *testimony* of others; by this means we can come into possession of a knowledge of facts which have never come under our own observation. Oral reports and written records furnish incomparably more information than any man's unaided observation could afford.

A further way of extending our knowledge is to be found in *inference*. From knowledge which we already possess we are able to arrive at conclusions which shall be true of things which may never have been observed by any one; we infer the cause of a distant sound, or the character of the other side of the moon, or the stature and habits of man's remote ancestors, or the climate of the Northern Hemisphere in the Carboniferous Age, etc. As we shall see later, *inference is involved in greater or less degree in all the other means to knowledge.*

An inference may, of course, be wrong; if it is to possess any degree of certainty, there must be a considerable body of information about the facts in question or about other facts closely related to them. The same is true, to a great extent, of memory, and even of perception, and to a very great extent in the case of testimony. Errors may arise at any point, and one of the most important problems in all thinking is the detection and elimination of errors.

Organizing Knowledge.—Classification as a Preliminary Step.—So far, attention has been fixed upon the processes employed in acquiring knowledge of facts.

In order to make this knowledge available, the data thus acquired must be arranged or classified. The object of science is to get organized knowledge, and before knowledge can be organized it must be so arranged as to enable us to see what facts are similar and what are different. *Classification* is the grouping of phenomena according to their likenesses and differences; those possessing a given characteristic are put into a group or class; those lacking it may be put into one or more other classes. Classes may be grouped together in a larger class or subdivided into smaller ones.

Language as a Necessary Instrument.—There is one very important instrument for the acquisition of knowledge which has not yet been mentioned; and that is *language.* Without some means of describing or otherwise representing facts, only a very limited use could be made of our perceptions: testimony would be impossible without it; inference involves representing to ourselves the consequences of certain principles or facts or situations; imagination and memory are ways of representing what is absent by means of pictures of the facts themselves or by means of other symbols. A great variety of symbols might be employed, but language, spoken and written, supplies by far the most important and complete set of symbols. The description and classification of facts would be practically impossible without language.

Further Steps in Organized Knowledge.—In some sciences we find little more than classified knowledge; the so-called "classificatory sciences," such as botany and zoology, have, until recently, consisted almost wholly of classified data. Science aims not simply at *classified* knowledge, but at *organized* knowledge, at knowledge organized into a *coherent system.*[5] It aims at the discovery of the *laws* manifested by its data as well as at the discovery of the data themselves and their arrangement into groups.

What is a scientific law? *A law in the field of science is a statement of the way in which things do invariably behave.* Unlike a moral law, a scientific law has nothing to say about the way in which things ought to behave, and, unlike a civil law, it does not prescribe a mode of action whose violation involves a penalty. The law of gravitation, for example, simply states that bodies do attract each other in certain definite ways; if bodies should fail to do this, the law of gravitation would be no genuine law. A scientific law states an invariable, unconditional connection between phenomena.

How are laws of this character discovered? They are based originally upon observation of particular instances of the behavior of phenomena. From observed instances we draw an inference which covers all other cases of the sort, past and future. This, that and the other acid turns blue litmus paper red; we conclude that all acids will have a like effect. This conclusion may, of course, be mistaken; it is an inference, and must be tested or verified.

Verification is then the next step. It may be undertaken in several ways: our conclusion may be com-

[5] There are of course fields of science where *classified* knowledge is the most that can be had. But the ideal of science goes beyond this.

pared with other things which we know about the facts under investigation; it may be shown to be a consequence of some known law; or it may be possible to find some further fact which would be consistent with our inference and with no alternative inference that can be suggested. Speaking generally, verification involves finding whether the inference in question fits in with the system of things to which it belongs. If such a test cannot be applied, if there is no such system of which it can be shown to be a member, it remains uncertain.

What is Presupposed?—One important question remains to be asked. Are there any laws or universal propositions which do not require verification? Are there any statements which are self-evident and not open to question or to proof? Axioms, such as those of mathematics, are sometimes said to be of this character. For example, take the statement that two things equal to the same thing are equal to each other; can this statement be doubted or can a proof for it be conceived? Are there not propositions which are so fundamental that they cannot be based upon any which are more general, and so necessary to all thought that they cannot be based upon perception, but are presupposed in perception? This raises again the question at issue between Locke and Kant; without attempting to answer it, we may at least say that no proposition which does not justify itself in experience can be accepted as true. Many propositions have seemed to be self-evident only to be proved false by later development in knowledge, and whatever else may be urged in favor of any proposition, it must at any rate fit in with the rest of the things we know if it is to be accepted as true.

Certain of these axioms or postulates are to be found in every science. In logic they appear under the name of the *Laws of Thought.* They are:

The Law of Identity, expressed by the formula: A is A.

The Law of Contradiction, expressed by the formula: A is not non-A.

The Law of Excluded Middle: Either A is B or A is not B.

The Law of Sufficient Reason: Every thing which exists has a sufficient reason or cause for being what it is.

There is some disagreement regarding the meaning of some of these laws. The Law of Identity, for example, seems to be a mere tautology: to state that A is A, or that a thing is what it is, does not seem to give us any information. It is true, of course, that in a world where the Law of Identity, in this sense, did not hold, reason could do nothing. But the Law of Identity is usually taken to mean also that there must be an element of identity in every act of thought and in every piece of reasoning. In the proposition "Man is rational," it is obvious that *man* and *rational* are not identical; still there is something common to the two; without this core of identity no single judgment would be possible.

The Law of Contradiction complements the Law of Identity. A thing is not its opposite, and in so far as there is opposition between two things it is necessary to assert that one is not the other.

The Law of Excluded Middle asserts that of two contradictory statements one or the other must be true. The law does not hold if the two statements are not contradictory, *i.e.*, if there is any third possibility. There is a middle ground between "A is brilliant" and "A is stupid": he may be an average person. But "This figure is square" and "This figure is not square" are contradictories.

All these laws are, of course, laws *for* thought, but they are equally laws *of* things, and they are laws for thought for that reason only. Certainly they must hold for any world in which reason can operate.

The Law of Sufficient Reason asserts that the universe is a rational universe; that for everything that exists there is a reason, and an adequate reason; that things are capable of explanation, implying that the world is a coherent system. In the words of Leibniz, who gave the principle its rank, ". . . nothing occurs for which one having sufficient knowledge might not be able to give a sufficient reason why it is as it is and not otherwise."[6] If the world were entirely chaotic, knowledge, except that of the most primitive sort, would be impossible; there could be no general knowledge, no knowledge of laws or principles, for laws and principles would not exist. It is conceivable, however, that the world is only partly rational, that there are things for which there is no sufficient reason; if so, rational knowledge would be limited to the fields within which principles did hold.

Summarizing, we may say that every science aims

[6] *Principes de la Nature et de la Grace.* Quoted in *Dictionary of Philosophy,* Ed. J. Mark Baldwin, Art. "Sufficient Reason."

at the discovery of the laws of the data with which it deals, and at the organization of all its content into a single systematic whole. A completely organized system of knowledge would be one in which every part would imply every other, and he who understood the system perfectly could reconstruct the whole from any part. Cuvier claimed that a naturalist could reconstruct an animal from a single bone, and he himself, as noted by Huxley in the passage quoted above, gave evidence of the validity of his claim. Perfection of organization is not to be found in any natural science; the mathematical sciences show something approximating completeness, but they do not deal directly with concrete facts.

CHAPTER II

FIRST STAGES IN KNOWLEDGE

I. Facts and the Ways in Which They are Known.—*Knowledge begins with the perception of facts;* and these facts are of many kinds. What is a *fact?* A fact is anything which exists; it is that which is real, apart from any opinion we may have about it or any attitude which we may take toward it; it is that which *is* as opposed to that which is merely imagined or conceived. When we ask for facts, we ask for something which shall be independent of any belief [1] or disbelief, approval or disapproval, on the part of any person.[2] Some or all of these characteristics belong to laws, but fact is distinguished from law in being concrete and particular, instead of abstract and general.

A. Perception and What It Includes.—Facts are known primarily through perception and memory; they are known *directly* only by means of perception,

[1] Belief and disbelief, whether true or false, are themselves facts; they are psychological facts. Belief in the Ptolemaic astronomy was a fact; that is, the belief actually existed. A false belief is one in which the thing believed is not a fact; it asserts or assents to something which does not really exist. Belief or disbelief may bring about changes in facts; in other words, give rise to new facts, as may any other existing thing. To say that a fact is independent of our attitude means that its existence and character are what they are apart from our attitude and aside from any possible effects which may be produced upon them by our attitude.

[2] This position is confessedly dogmatic. Further reflection might show that nothing is independent, but for our present purpose this position is justified.

though there are various ways in which they may be known *indirectly*. One of the most important of these indirect means, and one which is an important element in all the rest, is *inference;* a perceived fact may be evidence to our minds of the existence of something which we cannot perceive.

Much that is often included under perception must be eliminated when we are trying to use the term with scientific accuracy. For example, we say that we perceive the inkstand upon the table, or a man on the other side of the street, or that lightning has set fire to a distant building, or that Mr. X is an able lawyer, or that history repeats itself, and so on. Are any of these pure perceptions? We may perceive certain *events*, but to "perceive" that history is therein repeating itself involves, at the very least, these inferences: that the words of historians represent what has occurred in the past; that they are competent and truthful and that we understand them; and that the events we perceive are really like those which they have described. In the example of the lawyer, we base our belief on observation of certain acts of his which have brought about desired results in spite of difficulties; and on the inferences that he understood the situation and intended to bring about the results which actually occurred. Again, though the flash and the distant light were perceived, the conclusions that the flash was lightning and that the light was that of a burning building in the distance, and that the first of these was the cause of the second, involve far more than perception. In such instances as these the presence of *inference* is evident and the importance of distinguishing what is perceived from what is inferred is obvious. The perception might be correct, while the inference was erroneous, or vice versa. By distinguishing the two, the problems of discovering error and of correcting it are much simplified.

But it is by no means easy to know where to draw the line between perception and inference. We should say ordinarily that we perceive the ink-well or the man across the street, but even in these cases there is something which is very like inference. A perception contains many different elements, and these get themselves before the mind in a variety of ways; comparatively little in any perception can be said to come directly from the object. In the perception of the ink-well or the man all that we get directly is a spot of color with certain variations of light, shade, and so on. But we seem to see an object in three dimensions, of a certain size, at a given distance from us, and possessing weight, resistance, a certain degree of hardness, a peculiar internal structure and an indefinite number of other qualities, which may be more or less definitely present to the mind. If we had not, in the past, found these qualities in combination with spots of color similar to those now present, we should not be aware of them now; but that does not mean that these qualities are simply remembered, for they are present to the mind as *genuinely objective qualities*, and we seem to be as directly aware of them as we are of the color, although reflection shows us that they could not be given by sight alone. They all seem to present themselves together, while in remembering a number of events, first one appears before the mind and then another; in perceiving an object, the qualities do not come forward one after another, but all seem to be present together in a single thing. A perception is a *reaction* of the mind to an object, quality, or event of some kind. A mind which has had little experience in a given field will react to an object in that field with a perception of a comparatively simple sort; if one had seen and handled oranges but had not tasted them, his perception would contain no suggestion of the flavor, as a blind man's perception contains no suggestion of color or other visual qualities.

Every time an object is perceived under new conditions something is added which will modify future perceptions in greater or less degree. The child builds up his perceptions gradually; from a first vague, indefinite perception he advances to one that is more coherent and complete.

The way in which any person will perceive an object will depend largely upon his past experience: different persons will consequently perceive the same object differently; as no two persons have ever had precisely the same experience, they will never see a given object in precisely the same way. But in most instances the differences will be slight, because there is so much that is common in the experience of all, and in the perception of ordinary objects the differences are usually small and comparatively unimportant.

"Fallacies" of Perception, and Their Causes.—We think of perception as a certain and infallible source of knowledge; but if in all perceptions there is a large addition from past experience it is clear that many of them are likely to be wrong. The present object may

not be similar in all respects to like objects which we have seen in the past. A spot of color of a certain shape and apparent size may have stood invariably for an orange; in other words, it may have been found along with other sensations indicating a solid spherical object, of a certain flavor and odor, with a certain internal structure, and so on. If the spot of color again appears we seem to be aware of the other qualities. But there may be only a spot of color, as on the painter's canvas. Again, when two persons are similar in appearance, one may easily be mistaken for the other. The visual appearance of A may seem to assure the presence of the other qualities which, as a matter of fact, belong to B.

The possibility of erroneous perceptions was commented upon very early in the history of thought, and because errors of this sort occur so frequently, some thinkers concluded that the senses were altogether unreliable as sources of knowledge. Others urged that the fault was not with the senses; they pointed out that the trouble lay in adding to what the senses gave. When we have a sensation of greenness, they said, greenness is actually present to the mind; if we go on to say that there is present an apple, we are adding a number of qualities to those which are given by sensation, and the qualities we add may not really be there. If we should refrain from adding those other qualities we should never be mistaken, but it is impossible entirely to separate the sensational element from the others. The perception is a unit in spite of the complexity of the qualities which make it up, and these qualities are capable of modifying each other. The green of a picture or of a landscape does not look the same if the scene is looked at upside down; of course we should be correct in saying that we seem to see a certain shade of green in the first case and a different one in the second. We may be perfectly certain with regard to what we seem to see; but then we seem to see an object as having three dimensions, when it may have only two, as in a painting; what we usually want to know is whether we see the thing as it is, whether other people seem to see the same thing, whether we may expect to seem to see it in the future, whether handling the object would give confirmatory sensations, and so on. The attempt to limit our statements to what is unmistakably before the mind takes us a very little way toward certainty in knowledge. Perceptions include more than that. They should, of course, be made as carefully as possible. But although errors are certain to occur, yet if we do not run this risk of error we make little progress toward knowledge.

This first form then in which knowledge appears is open to mistake; there are mistaken perceptions. It may be well to note the different types of error and their chief causes. The types are usually said to be two; and they have been called **Mal-observation** and **Non-observation.** The names are self-explanatory. Mal-observation is of two kinds: in the one, something which does not belong to the object is added in the perception; in the other, the relations of the parts are wrongly perceived, as when we read *there* for *three.* Of course, both kinds of mal-observation may be present together, and non-observation also. Otherwise stated, there are really three kinds of error: *omission,* *addition* and *wrong relation of the parts in a whole.* They may occur at any stage in knowledge, and they are, in fact, the only kinds which can occur at any stage. What are their causes in the field of perception? A passage from Bacon, quoted by Jevons, calls attention to a number of the causes which give rise to them: "Things escape the senses because the object is not sufficient in quantity to strike the sense: as all minute bodies; because the percussion of the object is too great to be endured by the senses: as the form of the sun when looking directly at it in mid-day; because the time is not proportionate to actuate the sense: as the motion of a bullet in the air, or the quick circular motion of a fire-brand, which are too fast, or the hour hand of a common clock, which is too slow; from the distance of the object as to place: as the size of celestial bodies, and the size and nature of all distant bodies; from prepossession by another object: as one powerful smell renders other smells in the same room imperceptible; from the interruption of interposing bodies: as the internal parts of animals; and because the object is unfit to make an impression upon the sense: as the air, or the invisible and untangible spirit which is included in every living body."

The various kinds of causes may be classified as follows:

1. In the first place, the *external* or *physical conditions* of the perception may be unfavorable; in a red or green light, the color of objects is wrongly seen; in a fog, sounding objects seem nearer than they really are; if the light is dim, details are overlooked; if we look through an imperfect window-pane, objects ap-

pear distorted. In all these cases there is something in the medium through which the object is perceived which leads to error. Similar difficulties arise when instruments are employed to extend the range or increase the accuracy of our perceptions. Any imperfection in the instrument is almost certain to be a fruitful source of error. Other things might be cited in this field, but these will suffice to illustrate the class.

2. Next in order we may mention the *physiological causes* of mistaken perceptions. Imperfections in the sense-organ, fatigue, illness, and the like are obvious examples. There is one sort of perception which is always inaccurate, that of the time at which an event occurs; a flash of lightning is seen a fraction of a second *after* the light reaches our eyes; a sound is not heard in the instant at which it reaches our ears. The reason is this: a thing cannot be perceived until the nerve current which it sets up in our sense-organ has passed along through the nerves to the brain; this takes time, and in some cases, as in astronomy, the errors which arise from this source may be very important. Again, we often tend to perceive an event for a moment after it has ceased, since the nervous system continues to reverberate, as it were, after the original cause of its activity has ceased to act. Hence the flash or sound seems to be present after it has really passed. This is seen in our inability to distinguish the spokes of a rapidly revolving wheel or the successive vibrations of a tone, or single views in moving pictures; in all these the succeeding event begins before we have ceased to perceive the one before it.

3. But if all physiological and physical conditions

were favorable, if all organs and media and instruments were perfect, there would still remain the *psychological sources of error.* These are often or even always present along with the others. One of the psychological causes has already been alluded to, namely, (1) the tendency to see what we have previously seen in similar circumstances. There is also (2) a tendency to perceive what we expect or wish or hope or fear, or what has been recently or habitually in the mind, or that which has been vividly perceived or imagined. What is known as the "proof-reader's" illusion illustrates one of these; in reading, the context often suggests a certain word and we see that word and overlook mistakes in spelling. In the following passage (based on one in James's *Psychology*) few persons reading at the ordinary rate, and with ordinary care, would succeed in detecting all the mistakes in spelling: Any one wateing in a dark plase and expectng or faering a certaon objectt will interpret an abrup seusation to mean that object's presense. The boy playing "I spy," the criminel skulhing from his persuers, the superstitions personn hureying throuh the churchvard at midnight, the man losst in the woods, the girl who tremulusly has made an eveniug apointmnt with her swain, all are subjec to ilusions of sight and sound wkich make there hearts beat til they ate dispelld.

Another case illustrating some of these principles is that of the prisoner who had already been convicted of one crime and served his sentence, and who narrowly escaped conviction a second time, although entirely innocent in both cases. He bore a superficial resemblance to the real criminal; the witnesses were predisposed to believe that he was the criminal, and they positively identified him as the one whom they had seen committing the crime.

The effectiveness of all these tendencies is enhanced by (3) lack of attention or misplaced attention. The inattentive reader overlooks misprints, and so does the reader who is very intent upon the thought. Prestidigitators, fraudulent spiritualistic mediums, and the like, take advantage of these tendencies. They direct attention to unimportant things in order that they may do the important ones unobserved, and by leading the spectator to expect certain events they can often persuade him to believe that he actually witnesses them.

Mistakes as to the order of events are very easy in some circumstances; if two events, one in the field of sight and the other in that of sound, occur very nearly at the same time, this often happens. (4) Lack of training in observing events of a given kind may make correct perception impossible; the use of the microscope, finding and following a trail in the woods, seeing distant objects at sea or on the plains, distinguishing flavors, colors, etc., are examples.

(5) Abnormal psychological conditions, such as nervous excitement, those produced by drugs, etc., modify the keenness and accuracy of perceptions, sometimes for the better and sometimes for the worse. These various causes, physical, physiological and psychological, are so closely bound up together that it is often difficult to say which is chiefly operative in any given case.

Careful and intelligent attention will prevent many errors. A careful perception made with a purpose is called an *observation*. This term is sometimes used to cover all perception whatsoever, but it will be used here in the narrower sense. Still, the most carefully made perception may prove to be mistaken. In some of the sciences there are various special and technical methods of eliminating error.[3]

The discovery of error does not always lead to its elimination nor enable us to make the requisite correction. In some cases it does; we have already seen that the perception of an event takes time; this time is longer for some persons than for others, but for each it is approximately constant under given conditions. By means of a device which registers the exact time at which a certain event occurs and the time at which the observer indicates that he perceives it, it is possible to determine his "personal error" and to make the proper correction in cases in which the exact time of the occurrence cannot be otherwise determined. But in most cases it is not possible to do this or even to guess at the presence or the amount of error. Sometimes, as in the case of measurements, it may be possible to repeat the observation, and if the results cannot be made to agree, we can sometimes get a result approximately correct by taking an average.

Testing Perceptions.—It can almost be said that every observation should be held in suspicion until tested. The test would consist in finding out whether it agreed with other observations of the same fact or of similar facts made by ourselves or others, whether it was in agreement with the laws of the field in which it was found, with the laws of Nature generally, and so

3 See Jevons, *Principles of Science*, chap. xv.

on. We always proceed upon the principle that all knowledge should hang together, should be consistent and coherent; that the world is a consistent and coherent world; and that correct perceptions will agree with each other and with the rest of our experience.

When it is possible to repeat an observation, we have at once a starting-point for testing it; and when new observations can be made under more exact conditions, we are in a very favorable position for extending our knowledge of the facts under observation. One of the chief reasons why modern astronomy, for example, is so far in advance of that of the Greeks is to be found in the fact that modern instruments make the observations of astronomical phenomena so much more reliable.

Experiment.—Sometimes it is possible to reproduce at will the phenomenon under observation. This is the case, to a great degree, in physics and chemistry: sounds, chemical changes, and so on, can be repeated indefinitely. Moreover, the circumstances in which the phenomenon occurs may often be controlled and varied more or less, and that is very often a matter of great importance, as will appear later. *To bring about an event for the sake of observing it is to experiment.* An experiment may be performed for various reasons; it may be that we wish simply to get an additional observation as a basis for inference or a means of testing the accuracy of one which we have made already; or we may wish to see the result of changing certain of the circumstances in which the phenomenon occurs; or of finding the consequences of any condition whatsoever. "Whenever we can, by our own agency, influence the object we are investigating, we can remedy this want [insufficient observation] by experiment. We can institute at will a certain group of conditions C, and so compel the causes which are really at work to respond with an effect E, which would otherwise perhaps have never come within the domain of our senses. By varying at will the quantity and composition of that C we can bring about in E a series of changes in quantity and kind, which were still less likely to offer themselves unsolicited to our observation. Again, we can break up C into its component parts, and in each experiment allow but one of these, or a definitely assigned group of several of them, to take effect, at the same time cutting off the rest from action. The constituent elements of the result E admit of being separated in the same way, so that we learn which of them depends upon which element of the compound C. Thus experiment is the practical means by which we furnish ourselves with observations in such number and involving such mutual differences and affinities as is requisite in order to the elimination of what is unessential in them. . . . Defined in this way, it is clear that experiment only has an advantage over observation in so far as it is capable of supplementing the usual deficiencies of the latter; its function is to furnish us with suitable and fruitful observations instead of the unsuitable and unfruitful ones which offer themselves. . . . It is merely a way of preparing and setting before ourselves phenomena which it is of importance that we should observe."[4] But its function is exceedingly

[4] Lotze, *Logic*, Bk. II, chap. vii, 260.

important, and without it many sciences could make little progress.

The peculiar advantage of being able to control and vary the conditions of an event to be observed will be evident at a later point.

B. Indirect Means to Knowledge of Facts. I. *Memory and its Defects.*—So far in the present chapter we have discussed only the direct means of knowing facts. It appeared, however, that even in perception there is much that is a revival of past experience, reinstated by the memory. Knowledge of the past, reappearing in memory, bulks very large in the total of our knowledge. True memory is simply the recall of past experience accompanied by awareness of the fact that it was our experience. If one experience or one object of experience is similar to what is now before our minds, or if it has been related to the latter in any way, it tends to reappear. That tendency is often overcome, otherwise practically everything would be remembered. Not only do many things drop out of the memory, but many are also changed in their character or order, and some things may be added. Ordinary forgetfulness corresponds to non-observation. It is practically always present in greater or less degree, and it obviously tends to increase with the lapse of time. Many things disappear altogether; sometimes the main outlines are remembered and details forgotten; sometimes only a few of the details remain.

Remembering wrongly corresponds to mal-observation: words which were correctly heard may be incorrectly remembered; an object which was seen as red may be remembered as brown, and so on. Hardly ever do any two witnesses agree exactly in their memory of events which could easily have been observed with little danger of mistake. This is so generally recognized that too close a correspondence between the stories of two witnesses is regarded as an evidence of collusion and dishonesty.

Besides the modification of details, the order of events may be changed in memory or their relations may be modified in other ways, and entirely new elements may be introduced. Among the causes of mistaken memory the following may be noted:

1. A tendency to remember what would usually have happened in the circumstances.

2. A tendency to remember things or elements which were particularly pleasant or unpleasant, desired or feared, etc., at the expense of those which were more neutral. Elements or events of this sort, which did not occur but were suggested or expected, may be remembered as if they had occurred.

3. A tendency to remember things in a way which would make them more complete or logical, or more in agreement with our own opinions or wishes, or more in harmony with what we expected, or feared, etc.

4. Events which have often been described in one's hearing may seem to be *remembered*.

The tests of memory, like those of perception, are based upon the principle that genuine knowledge is always consistent and coherent, that the world of facts is throughout harmonious.

Where accurate records are available the memory, as a source of knowledge of the past, becomes much less important. Accurate records made at the time when the phenomena were perceived are an essential part of all the concrete sciences. The methods of recording are many, and they are too technical for present discussion.

II. *Testimony.*—Written records and oral reports make up a large part of what is known as evidence. Besides these, evidence includes historical remains of every sort, products of man's activity, natural phenomena of every kind, such as glacial scratches, geological deposits, etc., etc. The evidence may be of something in the remote past, of something not observed in the present, or of future events. The use of evidence clearly involves making inferences; it also involves perception. Some phenomenon is perceived, such as an uttered sound or an inscription or a fossil, and on the basis of this perception the observer draws conclusions concerning something which may never be perceived.

Oral and written reports, or, in other words, testimony, furnish a frequent ground of inference. Testimony includes every statement of fact made by any one. The opportunities for error in using it are so numerous that it is surprising that correct information can ever be reached by means of it.

1. In the first place, the person making the statement was liable to error in many ways when he observed the fact which his statement purports to represent.

2. In the second place, his memory is almost certainly inaccurate in one way or another.

3. Again, the words which he uses may not correctly represent to us what he has in mind; he may not use words accurately, or he may use them in a sense unfamiliar to his hearer.

4. In the fourth place, he may not be truthful; he may never have witnessed what he pretends to report, or he may intentionally misrepresent what he has witnessed.

These difficulties are present in both oral and written testimony; in the latter there are additional difficulties. What seems to be the witness of one person may be a garbled account; or errors may have been introduced by a copyist or an editor. In oral testimony cross-examination gives a basis for testing statements of the witness. In written testimony the substitute for this is found in other statements by the same writers and by contemporaries; when these are not to be found, little credence can be given to the testimony.

III. *Inference.*—Inferences from facts of every sort are also liable to error. In every case the final test is that of consistency and coherency. The application of the tests very often involves complicated reasoning and a large body of special information; it will be discussed incidentally in later chapters; much of scientific method is for the purpose of making such tests.

EXERCISES

1. How much is really observed in seeing a marksman shoot a clay pigeon? In hearing an automobile pass, a block away? In seeing a prestidigitator take an object from a pocket in which it was not?

2. What are the causes of mal-observation and non-observation in the following cases?

   (1) A straight stick partly immersed in water seems to be bent.

   (2) Two objects looked at through a stereoscope seem to be one, and they seem to be solid instead of flat.

   (3) The sun seen through a fog sometimes appears red.

(4) Mirrors increase the apparent size of a room.
(5) Distant objects appear small.
(6) Patients often seem to feel pain in amputated limbs.
(7) A table seems to throb if the fingers are pressed against it.
(8) A rearrangement of the furniture in a room is often unnoticed.
(9) We sometimes seem to feel the motion of a boat after landing.
(10) There are marked differences in what the ordinary good observer, the artist, and the botanist see in a flower.
(11) Silas Marner mistook Effie's hair for the lost gold.
(12) Looking at one's watch and not knowing the time a moment later.
(13) Not seeing the people one meets.
(14) In Poe's *Sphinx,* a small animal on the window pane is thought to be a large moth of a strange species.
(15) Mistaking the order of numbers, as 546 for 564.
(16) Finding a likeness between an infant and its parents.
(17) Macbeth seeing Birnam Wood coming to Dunsinane.
(18) The pain of amputation when, instead of amputation, an icicle is drawn across a limb.
(19) Shooting a man for a deer when hunting in the woods.
(20) The child's "seeing" things at night.

3. Give five examples of mistaken observation arising from each kind of cause described in the text.

4. Suggest causes for errors of memory in the following cases:

(1) Memory of "the good old days" as better than the present.
(2) Remembering the childhood of men who later became famous.

(3) In *Ivanhoe*, Wamba tells the travelers to go in one direction, but points in the other; one of them remembers the verbal directions, the other the direction pointed out.

(4) Forgetting cases which do not support one's view.

(5) Forgetting certain items in lists of things to be bought, etc.

(6) Dropping out characters or events in remembering a story or play.

(7) Ascribing to one person words or deeds of another.

(8) "Remembering" events which occurred before one was born.

(9) "Remembering" the apt replies which one might have made.

(10) Remembering as an actual experience what was merely a fiction often related as an experience.

5. In how many different ways could you account for the statement of a witness that he had seen a ghost?

6. Suppose three honest witnesses to have testified to seeing a man catch a bullet in his teeth: What would your conclusion be?

7. How would you test the statement: General X was killed in the battle of Gettysburg?

CHAPTER III

CLASSIFICATION

Objects of experience make their appearance in an order which seems to be almost chaotic; and in memory they are often reproduced very much in the order in which they originally occurred. But even in memory, and still more in reflection, there is a tendency to arrange things according to their likenesses and differences. This is the beginning of classification. Classification "is not identical with collection. It denotes the systematic association of kindred facts, the collection, not of all, but of relevant and crucial facts." [1]

A classification is necessarily based on a similarity of some sort: of quality or structure or origin, and so on. Any given collection of things may be classified in many different ways. Books, for example, may be grouped according to subject, size, style of binding, publisher, and so on; minerals, according to composition, value or chemical properties; the people of a city, according to race, income, occupation or religion. Any quality or relation whatever may serve as a basis of classification. In the abstract, one may be as good as another, and the one to be employed in a given instance will be that which best serves the purpose we have in hand. There are several different types of classification, each serving a special purpose.

Types of Classification.—1. Index Classification.—

[1] Karl Pearson, *The Grammar of Science*, chap. III., n. 1.

We may notice briefly the "Index Classification."[2] The purpose of this mode of grouping is to enable us to get hold of a given fact quickly and easily. Catalogues are usually constructed with this end in view, and they illustrate the principles involved. Certain obvious characteristics are selected, and very often a given item may appear under several different heads, as in cross-references. Alphabetical catalogues are the most familiar examples of the index classification.

2. Diagnostic Classification.—A second type is the "Diagnostic Classification"; its purpose is the identification of an object or the discovery of the group to which it belongs. "Nature Study" books abound in classifications of this sort. Here, too, certain obvious characteristics are made the basis of classification. Flowers, for example, may be classified according to color or time of appearance or habitat; or the main divisions may be made upon one basis, as color, the first subdivision on another, as time of appearance, etc. The identification of ailments by the physician depends upon a classification of symptoms made upon this plan.

3. "Natural" and "Artificial" Classifications.—Both index and diagnostic classifications are useful, but they do not, by themselves, lead directly to any greater knowledge of the facts or of their essential relations. They are based, for the greater part, upon superficial and easily noticed characteristics,[3] and have little relation to the essential properties of the things classified. It is often possible, however, so to group

[2] See Jevons, *Principles of Science,* chap. xxx.

[3] A diagnostic classification which is to be a sure means for the identification of any and all cases should be based on essential qualities. Those based on superficial and striking qualities serve

phenomena as to display at once their most significant characteristics. Compare, for example, the popular classification of the whale as a fish with the scientific classification of the same animal as a mammal. To call a whale a fish is to imply that it lives in the water, but tells little more; to call it a mammal tells us that it has warm blood, lungs instead of gills, a four-chambered heart, certain peculiarities of the skeleton, and so on.

Grouping data in such a way as to make manifest at once their essential characteristics is the aim of classification in science. Since science aims at complete and systematic knowledge, it will obviously select as the basis of classification in any given case that quality which does correlate the greatest amount of knowledge about the facts under consideration.

Scientists usually make a distinction between "artificial" and "natural" classifications. "It would be possible to classify all living things according to color, as white, yellow, green organisms, etc. Such a classification would, however, be artificial and destitute of scientific value because based upon a purely artificial and highly inconstant character. An interesting example of an artificial classification formerly employed is the system of Linnæus, who classified flowering plants into Monandria, Diandria, Triandria, Tetrandria, etc., according to the number of stamens. This was sufficiently convenient for a first rough arrangement, but was soon found to lead to the most incongruous association of plants agreeing in the number of stamens but differing in almost all characters. From such cases it is plain that plants and animals cannot be naturally classified

for ready identification of many cases, but not for all. See Bosanquet's *Logic* for a discussion of Diagnostic Classification as one based upon deeper and more essential qualities.

by likenesses or differences in a single character artificially selected. The entire organisms must be taken into account, and the natural classification differs from the artificial one in representing real relationship and not merely superficial likeness. Modern biology teaches that this relationship is of precisely the same nature as human relationship, *i.e.*, that it is due to community of descent from ancestral plants or animals. . . . The labor of determining the natural classification is much lightened by the fact that certain structures are often found as a matter of experience to be constantly associated or correlated, so that the presence of one indicates the presence of the others. In such cases a single character may be taken as the basis of a classification which is natural, because agreement in one character has been previously proved empirically to indicate agreement in the others. For example, it has been proved that the differences or resemblances of animals are correlated with corresponding differences or resemblances in their teeth. Hence mammals, to a great extent, can be classified according to the structure and disposition of the teeth. And so in many groups it is usually possible to discover empirically some one or few characters on which, by reason of their constant association with other characters, a natural classification can be based."[4]

Biology furnishes one of the best illustrations of a field in which a natural classification can be made, although even here in many cases there is no universal agreement as to which is the natural classification. For each of two characters or sets of characters might be correlated with a number of others, and it might be

4 Sedgwick and Wilson, *Biology*, p. 175.

difficult to decide which of the two correlated the greater number or the more important ones. Even if there were such agreement, it would not necessarily be permanent; new information might result in the selection of a new basis of relationship. The difference between natural and artificial classifications is, as Jevons points out, one of degree only: "It will be found almost impossible to arrange objects according to any circumstance without finding that some correlation of other circumstances is thus made apparent."

The principle employed in classification for scientific purposes is well stated in Huxley's definition, which was modified somewhat by Jevons and stated in the following form: "By the classification of any series of objects is meant the actual or ideal arrangement together of those things which are like and the separation of those which are unlike, the purpose of the arrangement being, primarily, to disclose the correlations or laws of union of properties and circumstances, and, secondarily, to facilitate the operations of the mind in clearly conceiving and retaining in memory the characters of the objects in question."

A scientific classification is ordinarily designed to serve the purposes here enumerated, but there may be cases, especially in everyday life, where our primary interest is not in getting a complete knowledge of things, but in getting together things which have a relation to some common purpose or problem; and in such cases the grouping together of things which are, in most respects, very dissimilar, may be justifiable. Custom-house regulations, for example, proverbially group together things which, apart from certain economic considerations, may be totally unlike. The so-called artificial classification may be entirely satisfactory as an index or diagnostic classification, though in a diagnostic classification, as we have seen, the use of essential qualities would furnish a surer means for identifying doubtful cases than the use of obvious qualities.

Classification, of whatever sort, is not simply bringing together data into a single group; it involves the further ordering of the data in sub-groups.

Division.—Breaking up the group into sub-groups is known in logic as "Division." The first thing to do in making a logical division is to select some characteristic which will serve to distinguish some members of the group from the rest. It may belong to some and not to others, or it may belong to all in different degrees, etc. The technical name of a character used for this purpose is *fundamentum divisionis*, or basis of division. The simplest sort of division is that in which the *fundamentum divisionis* is a character present in some members of the class and lacking in the rest. Material substances may be divided into those which are mineral and those which are not; or, to cite an ancient example, we may divide and subdivide substances as follows:[5]

Substance

Corporeal Incorporeal

Animate Inanimate

Sensible Insensible

Rational Irrational, etc.

[5] This is known as the Tree of Porphyry, so-named from the Greek logician who was the earliest writer to give a distinct account of this type of division.

Division of this kind is called *dichotomous* or *bifurcate*, from the fact that each group or sub-group is always divided into two.

A more complex classification results from selecting as the basis of division some character which is possessed by all the members of the class but with differences of degree or quality, etc. Books, for example, if classified according to subject matter, would fall into several groups, and each of these might again be subdivided into several more.[6] A dichotomous division

[6] As further examples of classification of this kind we may cite the following from Hyslop's *Logic,* pages 96, 97.

- Figures
  - Plane....
    - Rectilinear...
      - Trilateral
      - Quadrilateral
      - Multilateral
    - Curvilinear...
      - Circular
      - Elliptical
      - Parabolic
      - Hyperbolic
  - Solid....
    - Rectilinear...
      - Tetrahedrons
      - Pentahedrons
      - Cubes
      - Parallelopipeds, etc.
    - Curvilinear...
      - Spheres
      - Cones
      - Cylinders
      - Paraboloids

- Science.
  - Physical.
    - Mechanical...
      - Physics
      - Chemistry
    - Organic......
      - Biology
      - Physiology
  - Moral...
    - Political......
      - History
      - Sociology
    - Psychological.
      - Noetics
      - Aesthetics
      - Ethics

would, of course, be possible here; we might have something like the following:

Books

On history — Not on history

Not on history: On chemistry — Not on chemistry, etc.

But a dichotomous division would soon become unwieldly; moreover, it does not present the classes in such a way as to indicate which are coördinate. The example just given might seem to make history coördinate with all other subjects taken together, and chemistry might seem to be subordinate to history. It is desirable, usually, that the classification shall put coördinate classes on the same plane, and this the dichotomous division cannot do.

Moreover, this sort of division embodies very little information; it points out a class which has a certain character and another which lacks it; the latter is described in negative terms only. The other type of classification presents a number of classes, each described in terms of some positive quality. Still there are many cases in which interest centers in those members of a class which possess (or lack) some certain character. For some purposes the division of population into voters and non-voters, or into literate and illiterate, may be quite as satisfactory as any other.

Sometimes our information regarding a class is so imperfect that a dichotomous division is the only one we can use. In a given shipload of immigrants we might know that some were Italians, and know nothing about the nationality of the rest except that they were not Italians. Or it might be that those outside the class positively characterized had so little in common that no class or series of classes, coördinate with the first, could include them all: the chemist's division of *elements* into *metals* and *non-metals* illustrates this.

It has often been said that a dichotomous division is the only one which insures against the omission of any individual. Every member of a class must either possess a certain quality or be without it; all that are excluded from the first class are necessarily included in the second, while in the other sort of division it is easy to overlook something. But if many classes are to be included, the dichotomous division soon becomes almost unmanageable, and if it is not carried out to the end we will not discover every class, although all are formally included. If it is carried out exhaustively, everything will of course be identified; but the same would be true of any other type of classification. Jevons holds that diagnostic classifications should usually be dichotomous.

Requirements of Classification.—In any scientific classification (1) *the sub-classes must include all that is included in the main class*; and (2) *they must not overlap, i. e.*, no individual should belong to two classes at once. To classify *people* as *large* and *small* would violate the first of these rules, while to classify them as *large*, *small*, and *blue-eyed*, would violate both. Violation of the first rule results in incomplete division; violation of the second, in cross-division.

Incomplete division is a consequence of failure to carry out a division to the end; sometimes the principle of division does not seem to permit this, or some of the data included may be of so peculiar a character that they do not seem to fall into any well-marked classes. An escape from difficulties is sometimes found in a *miscellaneous* class, which shall include all cases not otherwise provided for. When this is employed, the classification is certain to include all cases. The *miscellaneous* class corresponds to the negative class in dichotomous division.

Cross-division is a consequence of employing more than one *fundamentum divisionis.* In the example above, both size and eye-color were employed as bases of division.

Every classification should be complete or exhaustive; it should provide a place for every item. But a sort of cross-division may sometimes be very useful, as in index or diagnostic classifications. Ordinary subject indexes, classification of books under author and subject, or of college courses under department and year, are cases in point, as is also the classification of a disease under each of its several symptoms.

A class which is divided into sub-classes is technically called a *genus;* while each of the sub-classes is a *species.* "Caucasian" is a *species* of the genus "man." If a sub-class were to be divided it would be a genus in relation to *its* sub-classes: Slavs are a species of the *genus* Caucasian. *Any class, then, regarded as inclusive of other classes is a genus, whereas if it is regarded as subordinate to some higher class it is a species.* A class which is so wide that no other can contain it is called a *summum genus* or highest genus; it alone can never be a species. A class which includes so little that it can not be subdivided is an *infima species* or lowest species; it can never become a genus.

An individual which is so unique that it can be included in no class whatever is *sui generis.*

Under ordinary conditions there is little use for these last three terms. It may be doubted whether there is such a thing as an individual thing *sui generis,* and whether there can be more than one *summum genus,* or any *infima species* which is not a class of one member only. In any given investigation they may be employed in a relative sense. For anthropology, mammal might be regarded as the *summum genus;* and an individual whose peculiarities defy all attempts at classification on usual lines might be spoken of as *sui generis;* a species which could not usefully be divided might be regarded as an *infima species.* But this use of the terms would not be entirely accurate.

The use of *genus* and *species* as described above is the traditional logical usage; but in the biological sciences they are used in a different sense. In those sciences the terms are not relative; a class is not a species at one time and a genus at another. *Homo* is always a genus; formerly it was thought to have two species, man and the chimpanzee, but now man, *homo sapiens,* is regarded as the only species in the genus. The Caucasian race is a *variety* under the *species, homo sapiens. Homo* is included in the *order, primates,* etc.

EXERCISES

1. What is classification?
2. What is an Index-classification? What is its purpose

and on what sort of quality is it based? How would you construct an Index-classification of the rulers of Europe during the Nineteenth Century?

3. What is a Diagnostic-classification? State its purpose and its principles. How would you construct a classification which would serve for the identification of birds?

4. What is the purpose of classification as used in scientific work? What is the difference between an artificial and a natural classification?

5. What is a Dichotomous Division, and what are its strong and weak points? Make a dichotomous division of educational institutions. What is a cross-division? How is it caused, and when may it be useful? Give examples of a genus, a species, a *summum genus,* an *infima species,* a thing *sui generis.* Contrast the biologist's use of "genus" and "species" with the more general logical usage.

6. Criticise the following classifications and divisions:

a Men may be classified as white and colored.
b Trees, as fruit-trees, shade-trees and forest-trees.
c The fine arts, as sculpture, painting, drawing, architecture, poetry and photography. (Fowler.)
d Books, as those on history, science, poetry, religion and belles-lettres.
e Political parties, as conservative and radical.
f The states of New England, as Maine, New Hampshire, Vermont, and Connecticut.
g Mind, into intellect, feeling and will.
h Body, into extension, weight, resistance, etc. (Mellone.)
i Religious, into monotheistic and polytheistic.
j Americans, into white, black and foreign-born.
k Politicians, into honest and dishonest.
l Books, into dull and interesting.
m Games, into those which are athletic and those which are intellectual.
n Pictures, into paintings, engravings, posters and pen and ink sketches.
o Domestic animals, into those which are useful and those which are pets.
p Motion, into molecular and molar.
q Bodies, into light, heavy, and dense.

r Men, into those whose main pre-occupation is to get through time and those whose aim it is to find time for all that has to be got through.

Can you state circumstances in which any of the above might be useful and satisfactory?

7. Divide and sub-divide: Propositions, Athletic sports, College publications, Government, Poetry, Furniture, Races, Schools.

8. Criticise the classing together of negroes, coal, and black chalk on the ground that they are similar in being *black, solid, extended, divisible, heavy,* etc. (Mellone.)

CHAPTER IV

THE USE AND MISUSE OF WORDS

Discrimination, Conception, Abstraction.—It will be remembered that a thing is put into a given class by virtue of its possession of some quality or relation, a class being simply a group of things which have in common one or more qualities or relations. Any given thing might, therefore, be classified in several different ways. Bucephalus, for example, might be classified as a horse, or as a colored object, or as a consumer of hay, or as a possession of Alexander the Great, and so on. Indeed, most concrete objects might be classified in hundreds of ways. For every characteristic which a thing possesses there may be a class, and the way in which we shall classify it in any given instance will depend upon the purpose we have in view. For his teacher the small boy is a pupil; for the cat, a source of danger, and so on. And each mode of classification is correct in its place.

But before an object can be classified in a given way it is, of course, necessary to note what qualities it does possess. Ordinarily we note very few of these. Most of us see only the most obvious and striking qualities of things, and we often see those very imperfectly. We get a vague general impression and fail to analyse it into its elements. The child, in his earliest experiences, hardly discriminates the different qualities of a thing at all, for his first experiences are very much confused.

We know things only as we know their qualities and relations, and the better we can distinguish and relate these the better we know the object. *Analysis* of the concrete datum is presupposed in classification and in all the other higher manifestations of consciousness. When we analyse a thing we pick out its various elements and think of them as more or less isolated from the complex in which they were perceived. We can think of greenness or roundness without thinking of size or hardness or of any of the other qualities with which greenness or hardness always occurs. We never *perceive* greenness or hardness by themselves, but we can *think of* them without taking into account the other qualities. The mental act whereby we think of them in that way cannot be perception; nor can it be memory, for memory, like perception, is of concrete complex things, whereas these qualities are simple and abstract. The mental act in which we bring before ourselves a simple quality is *conception;* and the thought of the quality is a *concept.* We have concepts of *abstract qualities,* but we cannot have percepts of them. But we may also have concepts of *concrete* things. Our idea of some particular horse, say Bucephalus, is not a memory, nor is it a perception; it is a concept.[1] We may also have concepts of *things which we have perceived;* indeed, every perception involves conception as well. We are immediately aware of certain qualities, but, more than that, we have an idea of a complex whole possessing more or less coherence, permanence, etc. Whenever we think of a *class of objects,* qualities, or

[1] Both *Concept* and *Conception* are used for the idea of the thing or quality.

what not, we do so by means of conception. The thought of *anything* is a concept. Some concepts are universal, some are particular; some are concrete, some are abstract; some are of real things, some of imaginary things, and so on. Everything that is thought of is thought of in a concept, or rather, the thought of anything is a concept of that thing.

In conceiving anything two elements are present: the *symbol*, and its *meaning*. In the concept "horse" the symbol is the word "horse"; the meaning is the sum-total of qualities which that word implies or the objects to which it may be applied. The symbol is not necessarily a word: we might think of horses without having in mind the word. The mental picture of a horse might be the symbol. If we were trying to convey the idea of a horse to a person who did not understand English, we might use a drawing or imitate the sound of galloping, and so on. In all these cases the meaning would be the same, though the symbol would not. The essential element in the concept is the meaning; so long as that remains the same we have the same concept, no matter what the symbol. The same thoughts may be present in two minds, one of which thinks in English and the other in German, or one of which thinks in words and the other in mental pictures. The superiority of words as symbols will be discussed presently.

It is customary to treat logic as if it dealt solely with concepts, judgments and inferences; but in treating logic as a part of scientific method, as a part of the science of getting knowledge, it will be well to continue to speak as if we were dealing directly with the facts and not with mental counters. In other words, logic may be regarded as a science of things as well as a science of thoughts. It deals, it is true, only with the most general aspects of things; not with their special qualities, as do the special sciences, such as physics, etc. It has to do with that which is common to all fields of facts. In certain cases it may be more convenient to speak of the concepts rather than of the things conceived, as, for example, in geometry, where the things conceived are certain highly abstract relations and the like; but even in such cases the other way of speaking would be possible.

Necessity for Language.—Mention has already been made of the necessity for describing or in some way representing the things we know. The means most universally employed and most completely developed is, of course, language. A language, from the point of view of logic and scientific method, is simply a highly complex system of symbols for the representation of all kinds of objects and experiences, of the conclusions and constructions based upon experience, of laws, and so on. Language, as already noted, is a condition of all progress beyond the merest rudiments of knowledge. It might seem to be of no use in the field of *observation*, but an observation made for some special purpose or under experimental conditions implies a previous statement or representation of the thing to be observed. *Memory* includes a representation of past experience by means either of a mental picture or of some other sort of symbol, such as the name of the thing. All spoken and written *evidence*, of course, implies language; and inference involves the statement to ourselves of a conclusion from something observed or thought of. *Classification* obviously requires the use of symbols.[2] So important is language for the work of thinking that logic has sometimes been defined as a branch of the study of language. Whately said that "Logic is entirely conversant about language." Some have maintained that all growth in thought has followed the development of language and would have been impossible without it; in other words, that language always precedes thought; that man is intelligent because he has language and not *vice versa.* These may be extreme views, but it is certain that systematic knowledge cannot go far without a coherent system of symbols, and that language is infinitely superior to all other kinds of symbols.[3] Any examination of the processes by which knowledge is attained must give careful attention to the consideration of language.

If language were perfect, it would not be necessary to discuss it at any length in this connection, but its imperfections are such as to lead very often to mistaken ideas and wrong conclusions. It will be necessary, therefore, to examine language in order to discover these imperfections and the means of avoiding them.

Terms.—A word or group of words stands as the representative of some thing, quality, relation, action,

[2] Sometimes the mental picture of an individual may stand for the class; the image of a tree may stand for the class *tree*, but when we know the name of the class or kind, we usually represent the class to ourselves by means of the name. Mental pictures are liable to vagueness and to modification, and it has been shown that scientists, for example, tend to represent things to themselves almost exclusively by means of words, particularly as they advance in years.

[3] For a discussion of this question see Stout, *Manual of Psychology,* chap. v.

idea, and so on, or some group or combination of them. Such a word or group of words is called a *term.* A term might consist of any number of words and contain various subordinate clauses, but if it stood as the symbol of some single object of thought it would still be one term. "Man" and "The torch in the hand of the Statue of Liberty in New York Harbor" are equally terms, for each stands for a single object of thought. The inevitable *difficulties with regard to the use* of terms arise from the fact that a word or a group of words may stand for more than one thing; it may have a variety of meanings and is therefore liable to misinterpretation. Most words are used in more senses than one, so the danger of confusion is always present. There are several causes for this multiplication of meanings. One of the most important is (1) the tendency to use a word in a sense wider than the one in which it was formerly used, to use it more generally, to generalize it. *Lens* meant originally only a double convex piece of glass; such words as *curve, acid, metal, salt,* etc., illustrate this tendency.

There is also (2) a contrary tendency to limit the application of words, to use a term in a narrower sense than formerly, a tendency toward specialization. *Minister* meant originally a servant; now it means, among other things, the highest representative of a state, one of the most exalted "servants" of a government. "*Deacon, bishop, clerk, queen, captain, general,* are all words which have undergone a like process of specialization. In such words as telegraph, rail, signal, station, and many other words arising from new inventions, we may trace the progress of change in a lifetime." The use of *Congressman* to describe Representatives only, and of *Protection* as a name for an economic policy, are further illustrations of the process.

These tendencies may affect a word and its derivatives in different degrees and different directions. Compare, for example, *distinguish* and *distinguished; dissolve, dissolute* and *dissolution; matter* and *materialistic; respond* and *responsible; design, designer* and *designing*, etc. The popular and the technical uses of a term are usually different, one being broader or narrower than the other; *phenomenon, sensation, idea*, are illustrations. Sometimes a special use of a word is only local, as in dialects, or for a short space of time, as in slang.

In addition to these tendencies to generalize and to specialize the meaning of terms there is (3) another by which there is a transfer of meaning to associated objects or to those which are analogous. The use of the word *church* to designate a religious society, or of *chair* to indicate a presiding officer, or of *bench* to stand for the judiciary, are illustrations of the transfer of meaning to associated objects, and such expressions as a *dull* student, a *hard* examination, a *brilliant* game, etc., illustrate the transfer to analogous objects.

There are (4) some other cases of less importance: sometimes two words which were originally different and of different derivations are alike in sound and spelling and might possibly be mistaken for each other; such as, *mean* in the sense of *middle* and *mean* in the sense of *low; pound* in the sense of *weight* and *pound* meaning a *pen, pen* as an inclosure and *pen* as an instrument of writing, etc. Sometimes words are alike in sound but different in spelling, as *right*, *wright* and *rite*, or *rain* and *reign;* in other cases they are alike in spelling but different in sound, as *lead*, the metal, and *lead*, something to be followed, etc.[4] These last three cases are of little importance since the confusions resulting are usually of only momentary duration. In the first three cases, those of *generalization*, *specialization* and *transfer of meaning*, the confusion results from the continued use of a word in the older sense after its meaning has been extended to a new field.

If the various meanings are clearly distinguished and widely separate, the context will usually make clear which is intended; but the meanings are most frequently very similar or closely connected, otherwise the same word would never have been used for both. The serious consequences of these confusions are seen in the fact that so many disputes and differences of opinion result from a difference in the use of terms.

Kinds of Terms.—1. SINGULAR AND GENERAL.—We have discussed the causes of ambiguity in terms; it will be well to examine some of the different kinds of terms in order to discover just what sorts of confusion are likely to be found. There are some words, such as proper names, which would not seem liable to ambiguity since they usually have little or no *meaning;* but after all there may be uncertainty enough with regard to the *application* of the name, as every case of mistaken identity shows.

Proper names are simply one variety of INDIVIDUAL

[4] Most of the illustrations used in the two pages above have been taken from Jevons, *Lesson in Logic.*

or SINGULAR terms. "The first president of the United States" is quite as definite in its application as any proper name could be. A singular term is a term which can be applied, in a given sense, to one single, individual object only. On the other hand there are terms which may be applied in the same sense to an indefinite number of objects. Man, president, book, college, etc., are GENERAL terms. A term which was originally singular may become general, as illustrated in the expressions, "A Daniel come to judgment," "A Homer," "A Hannibal," etc. On the other hand singular terms may be so combined as to apply to only one individual; the first president, the wisest man in the world, the longest river, the highest mountain, etc., are such cases.

The chief difficulty in the case of singular terms is the liability to error with regard to the individual to whom the name applies. The form or the context usually shows clearly enough whether a term is singular or general.

In the case of general terms there is much more difficulty: in the first place the meaning may be *vague;* the application of the term may not be definite; in the case of such words as *rich* and *poor, wise* and *ignorant,* and a great many others, there is no universal agreement and there is seldom a definite notion as to the range of application in any particular case. Again, where the term has a plurality of applications, each of which may be sufficiently definite, the wrong one may be employed or understood in any given case. The word *law,* for example, means, in one field, a prescribed rule of action, something imposed from without and having a binding force, as a civil law. In the natural sciences, on the other hand, a law is simply a statement of the way in which things do invariably behave. Obviously, one who carries over into the consideration of natural phenomena the conception of law employed in legal practice, is liable to have a very mistaken view of Nature. There is a multitude of terms in which such differences are to be found. The Latin *lex* meant originally something fixed or set, so both these meanings might be regarded as specializations in different directions of the original meanings. In other cases one meaning is obviously more or less general than the other. In the commandment, "Thou shalt not kill," the word *kill* is obviously less general than is the statement, "To kill is to deprive of life"; or again, the words *rest*, *sleep*, etc., are not intended to cover all possible cases in the statement, "Rest, food and sleep are necessary to life." Any mistake as to the exact sense in which a word is used will be certain to lead to mistaken opinions.

2. Concrete and Abstract Terms.—This brings us to the distinction between *concrete* and *abstract terms*, or between terms used in an abstract sense and the same terms as used in a concrete sense.

An abstract term, as the name implies, stands for something which is the product of abstraction; it is something separated from its context and considered by itself. For example, qualities are known only as they occur in an object, in a complex something which includes many other qualities. We have already seen that these qualities are not recognized as separate things in the child's earliest experience.

If a quality always appeared in the same setting, it would never be discriminated and hence could never be abstracted. To quote Professor James's illustration, if all wet things were cold and all cold things were wet, we should never distinguish coldness from wetness. But most qualities do occur in a variety of settings and can therefore be discriminated and abstracted. When a quality is abstracted it can be treated, for certain purposes, as a separate thing. In studying color, we disregard the other properties of colored objects and treat their colors as something independent.

There are many degrees of abstraction: blue is abstract as related to a blue object, but comparatively concrete as related to color. An abstract term may be the name of a relation, as height, or an action, as walking, or of any characteristic whatever, abstracted from its setting and regarded as an independent thing. The word which stands for this characteristic will be an abstract term. A given term is often used in both senses: in the sentence, "Government is necessary to civilization," the term government is said to be used in an abstract sense; in the sentence, "This government is a republic," it is concrete. To confuse the more general with the less general meaning of a term or the abstract with the concrete use of it, or to argue from a term taken without qualification to the same term qualified in some particular way, is to commit a fallacy, the *Fallacy of Accident.* To conclude that it would be meritorious to give a beggar a dollar because charity is a virtue would be to commit this fallacy. To conclude that, because the only Filipinos you have seen are small, a Filipino is a small person, would be to commit what is sometimes called the *Converse Fallacy of Accident;* in this, we argue from the concrete or the less general or what is true in particular circumstances, to the abstract or the more general or to what is true apart from particular circumstances.

The ancient example, "What is bought in the market is eaten; raw meat is bought in the market; therefore raw meat is eaten," illustrates the simple fallacy of Accident. So also does the following: "The Greeks produced masterpieces of art, and as the Spartans were Greeks, they produced masterpieces of art." (Davis.) "Greeks," in the major premise, is used in the generic sense. In the minor, it has a more specific meaning.

To argue that strychnine should be freely sold because it is very useful (as a medicine) would be to commit the converse fallacy.

3. Collective and Distributive Terms.—Another distinction which is of great importance in dealing with terms is that between collective and distributive terms. *Army,* for example, is a collective term; it stands for a number of individuals taken together in a group; it is a group term. A term like *man,* on the other hand, has no such significance. It applies equally to any and every individual in a class. In an expression like *all men,* for example, there is danger of confusion; it might be taken to mean all taken together, as in "All living men number about 1,500,000,000"; or it might be used distributively, as in "All men are mortal." Synonyms of the terms *collective* and *distributive* are *jointly* and *severally.* Obligations are sometimes laid upon individuals which they are bound jointly or severally to observe.

Confusion between the collective and the distributive uses of a term leads to the *Fallacies of Composition* and *Division;* arguing from the distributive to the collective use results in the fallacy of Composition. "Each member of the committee is insufficiently informed, therefore the committee as a whole is not sufficiently informed," contains a fallacy of Composition. But to argue that because a navy as a whole was weak, the individual ships were therefore weak, would be to commit a fallacy of Division. These fallacies should be kept clearly distinct from the fallacy of Accident. Here we are dealing with a group; the question is, are we dealing with it simply as a group, or are we thinking of the individuals of which it is made up? In the other case we were not concerned with a group at all.

4. Other Kinds of Terms.—Various other distinctions might be made among terms; there is, for example, a distinction between *positive* and *negative* terms; the former being those which imply the presence, and the latter those which imply the absence, of a quality. *White*, *just*, *warm*, etc., illustrate the former; *blind*, *empty*, *unconscious*, the latter. There is little danger of confusion in this and in most of the other cases which might be included, so we will pursue them no further.

Definition.—In any case in which misunderstanding is likely to occur, the first thing to do is, obviously, to make clear what it is that the term stands for. In the case of a proper name the application of the term could be shown by producing, or pointing out, or describing, the individual thing for which it stands, and so of other singular terms. But with general terms that is not possible; a general term stands for all possible cases of a given sort, past, present and to come, and any example or series of examples could at most *illustrate* the meaning of the term. Sometimes an example will show clearly enough, for ordinary purposes, what is meant. But any example might illustrate a variety of things; if two persons, each of whom was entirely unacquainted with the language of the other, should try to communicate by pointing to objects to indicate the meaning of the words they were using, they would illustrate the uncertainty of this method in its extreme form. If one of them should point to a horse, he might mean any one of a dozen different things: horse, or simply animal, or useful animal, or large object, or gray, or beautiful, or dangerous, and so on. In a minor degree that sort of difficulty is always present when illustrations are employed to indicate the meaning of terms, and the method of illustrations is never entirely satisfactory.

In Plato's *Euthyphro* Socrates asks Euthyphro, who claims to have a precise knowledge of the subject, "What is piety and what is impiety?" The reply is, "Piety is doing as I am doing; that is to say, prosecuting any one who is guilty of murder, sacrilege, or of any other crime—whether he be your father or mother or some other person, that makes no difference;—and not prosecuting them is impiety." But Socrates is not satisfied with this. "Remember," he says, "that I did not ask you to give me two or three examples of piety, but to explain the general idea which makes all pious things to be pious." In other words, what quality must things possess in order to be called pious? When we ask for a definition of a term we wish to know what qualities a thing must have in order to make the term applicable.

It should be noted that nearly all terms have two aspects: they *stand for objects* and they *imply* the *qualities* which those objects possess. The term *man* stands for any or all individual men, past, present and to come. It also implies all the qualities which a being must possess in order to be included in the class. In the case of a general term it is obvious that all the individuals for which it stands—in technical language, its total *extension*—could never be presented. The only way of indicating the full extent of its application is to show what qualities it implies, to tell what its *intension* is. That might conceivably be done by enumerating all the qualities which were regarded as essential. If things in a given group were so unique that they could not be included in a larger class, enumeration of their qualities would be the only way of showing what they were. But in all ordinary circumstances, this process can be abbreviated by stating (1) the class to which the things belong and (2) the quality which distinguishes them from the other members of the class.[5] The class-name will obviously imply the presence of

5 "Definition" has been variously defined. "Given any set of notions, a term is definable by means of these notions when, and only when, it is the only term having to certain of these notions a certain relation which itself is one of the said notions." (Russell, *The Principles of Mathematics*, Vol. I, chap. xi, sec. 108.) A term is defined by being given a place in a set of notions, which place can be occupied by it and by it alone. Instead of being assigned to a class it is given its place in a complex system of concepts. This last might be regarded as the more complete form of definition, whereas the former, the traditional form, is less complete, though adequate for ordinary purposes.

the qualities which these things have in common with the others in the class.

The class which includes a thing is its *genus:* "plane figure" is the genus of "triangle." And the quality which distinguishes it from the other members of the class is its *differentia* or *peculiar property;* "three-sided" is the differentia of "triangle." A *property* is any essential quality: having an equal number of sides and angles would be a property of "triangle." An *accident* is a quality which may or may not be present in any or all members of a group: having a right angle is an accident of "triangle."[6].

Defects of Definitions.—There are several defects to which definitions are liable.

1. They may be too broad, *i. e.*, they may include more than the term is intended to cover. To define a square as a rectilinear figure would be a case in point. In such definitions the differentia is not given or not properly given.

2. Again, the definition may be too narrow, that is, it may exclude part of what the term is intended to cover. To define "American citizen" as one born in the United States would exclude naturalized citizens. In this case an accidental quality is taken as the differentia.

3. Definitions are sometimes given in obscure or figurative or ambiguous language. Dr. Johnson's definition of a network as "anything decussated or reticulated with interstices between the intersections" is a

[6] These four terms, *genus, differentia, property* and *accident,* together with *species,* are what have been traditionally known in logic as the five Heads of Predicables or the five ways in which a predicate may be affirmed of a subject.

favorite illustration of the obscure definition, the definition of *ignotum per ignotius*. Spencer's definition of evolution as "an integration of matter and a concommitant dissipation of motion; during which the matter passes from an indefinite incoherent homogeneity to a definite coherent heterogeneity; and during which the retained motion undergoes a parallel transformation," is sometimes charged with this fault. It should be remembered, however, that when a term is to be used with scientific exactness it may be necessary to couch its definition in very technical terms; to one who had read the discussions which lead up to it, Spencer's definition would not seem obscure. Figurative and ambiguous language should always be avoided when exactness is the aim. Such language may give some suggestion of the meaning of a term, but does not really define it.

4. Sometimes unessential attributes are employed in defining a term: *e. g.*, "Books are the things out of which libraries are made." Such definitions are obviously faulty. They use an accident as the differentia and do not give the *meaning* of the term.

5. Whenever it is possible, a definition should be stated in positive rather than negative terms; to define "an under-classman" as "a student who is not an upper-classman" is to tell what he is not instead of telling what he is.

6. Another sort of bad definition is one in which the definition contains the term to be defined or some synonym or exact correlative of it. "Life is that which distinguishes living from non-living things" would be a flagrant case; "A cause is that which produces an effect" is little better. A definition should state clearly

the exact meaning of the term to be defined. There are cases, however, where a complete definition is not necessary. Where the hearer is in doubt which of two well-known meanings is intended, or where the term is already familiar but is used with a slightly different shade of meaning, in other words, where the genus is already known, the briefest indication of the differentia may suffice; sometimes the mention of any accidental quality, or even the use of an illustration, may be sufficient.

EXERCISES

1. Make a list of ten words which are sometimes misused through the fact that they have undergone generalization; a similar list of those which have undergone specialization; a list of five in which there has been a transfer of meaning to analagous objects.

2. Bring ten examples of singular terms and ten of general terms; five of general terms which were originally singular or were derived from singular terms.

3. Give ten examples of collective terms and ten of distributive. Show how error might arise in this connection.

4. What is a definition? Compare definition and description. Define the following terms: Book, Party, College, Republican, Honesty, Foot-ball, Dormitory, College-spirit, Club, Money, Success, Trustee, Tariff, Saint, Geometrical figure.

5. Criticise the following definitions:

(1) A phonograph is a mechanism for recording and reproducing sounds.

(2) A sea is a body of water, next in size to the oceans, entirely, or almost entirely, surrounded by land.

(3) A library is a collection of books generally for personal use and not meant for merchandise.

(4) A wagon is a conveyance mounted on wheels and drawn by some animal, usually a horse.

(5) Oxygen is the most important gaseous element known, without which combustion and animal life would be impossible.
(6) A sensation is a modification of consciousness produced by the excitation of a cortical center through the agency of an afferent nerve-current.
(7) " Life is a continuous adjustment of internal to external relations."
(8) Logic is the Baedeker of the world of thought.
(9) A cause is that which produces an effect.
(10) A book is a combination of leaves and cover.
(11) A sun-dial is an affair for telling time by means of the sun.
(12) Public opinion is the opinion of people generally.
(13) A student is one whose principal business is study.
(14) A just judge is one who never shows partiality in his decisions.
(15) Wood is the ligneous part of trees.
(16) Football is a game which is usually played in America with a large ball in the shape of an oblate spheroid, whereas in England a spherical ball is used.
(17) A liar is a man who wilfully misplaces his ontological predicates.
(18) A philosophical work is one which treats of some metaphysical subject.
(19) A philosophical work is one which deals with something abstract and difficult.
(20) A false weight is an abomination.
(21) The quality of a proposition is what tells us whether it is affirmative or negative.
(22) Definition is telling what a word means.
(23) A religion is that which satisfies the highest needs of man.
(24) Matter is the stuff out of which things are made.

6. What fallacies are committed in the following cases?

(1) The holder of some shares in a lottery is sure to gain a prize, and as I am the holder of some shares in a lottery I am sure to gain a prize. (Hyslop.)

(2) A monopoly of the sugar-refining business is beneficial to the sugar-refiners; and of the corn trade to the corn growers; and of silk-manufacture to the silk weavers; and of labor to the laborers. Now all these classes of men make up the whole community. Therefore a system of restrictions upon competition is beneficial to the community. (Hyslop.)

(3) Who is most hungry eats most; who eats least is most hungry; therefore who eats least eats most.

(4) All the trees in the field make a dense shade; therefore this elm tree, which is one of them, makes a thick shade.

(5) Cities are governed by mayors; hence a mayor was the highest official in ancient Rome.

(6) The major received a D.S.O. for attacking the enemy and appropriating their supplies; therefore it is praiseworthy to steal.

(7) The Irish are quick-witted; hence that Irish policeman must be quick-witted.

(8) This ship is one of the best in the world, for it belongs to the British Navy, which is the best in the world.

(9) Americans are liberal; hence this man may be counted on to give liberally, since he is an American.

(10) We can now see the results of giving the negro all the rights and privileges of the white man. Two months after he was placed in office, this colored man absconded with all the funds under his control.

(11) Every man has a right to teach his religious beliefs; therefore it is not out of place for a college instructor to do so in the discharge of his duties.

(12) Any student in college would stand higher in his class if he received higher marks; hence if all marks were raised 10% every man would stand nearer the head of his class.

(13) Pine wood is good for lumber; matches are pine wood; therefore matches are good for lumber. (Hyslop.)

(14) To teach a child is to improve him; showing him how to pick pockets is teaching him; hence that improves him.

(15) Poisons cause death; nux vomica is a poison; therefore it causes death.

(16) This reformer was working for selfish ends all the time; no more reformers for me.

(17) Since attending that socialist meeting I have had no confidence in socialistic doctrines.

(18) He cannot be innocent, for he was a member of the mob which committed the deed.

(19) Those two horses would make an excellent team, for each is the best of its class.

(20) Five is an odd number; three and two are five; and hence each is an odd number.

CHAPTER V

PROPOSITIONS

Difficulties in the use of language are not all provided against by the correct definition of terms. Many arise in the combination of words into sentences. A term, as we have seen, is the representative in language of some object of thought, real or imaginary, concrete or abstract. But the mind never rests in the contemplation of a single object; it always tends to make an assertion or judgment about this object. Most logicians are now of the opinion that, even in the simplest perception, a judgment is either present or implied. Introspection will show at once that when we hold an object before the mind, there is an inevitable tendency to think some assertion about it. The expression of this mental assertion or judgment in language is a proposition.

Kinds of Propositions.—Propositions are usually distinguished according to *quality* and *quantity*. (1) The qualities are two, *affirmative* and *negative*. The difference between affirmative and negative propositions is sufficiently familiar. It should be remembered, however, that the mere occurrence of *not* or some other negative particle in a proposition does not necessarily make the *proposition* negative. The proposition, "Those who do not study are in danger of failing," is not a negative proposition. It asserts positively something about a certain class, namely, "those who do not study"; these words constitute a negative *term*. An affirmative statement can be made about a negative subject as readily as about any other. In the proposition, "Those who do not study are unwise," the term unwise is also negative, but the proposition is affirmative. To decide whether any proposition is affirmative or negative, determine whether something is affirmed or denied of a subject. What the subject is, and what the predicate is, makes no difference; the only question is, do we affirm something or do we deny something? (2) With regard to quantity, propositions may be either *universal* or *particular*. A universal proposition is one which expresses a judgment about the whole of the class to which the subject applies. "All the stars are suns" is a *universal* proposition; so is "No planets are self-luminous." (The latter proposition is negative and denies something of all planets.) "Some stars are double" is called a *particular* proposition. It asserts something of some individuals of the class "stars." By "particular proposition" is not meant a statement about some particular individual. The proposition "Jupiter is the largest of the planets" is not a particular proposition. It is a *singular* proposition, but, since it expresses a judgment about the whole of that for which the term "Jupiter" stands, it may be treated as a universal proposition. The so-called particular propositions are really *indefinite;* if the "some" in any proposition meant certain particular ones, as it does in certain cases, the proposition would really be universal; it would say something about all those of whom the assertion was made: as "*some* persons" (meaning A, B, C) "are certain to be late." As used ordinarily, *some* means *certain unspecified individuals, it may or may not be all.* The word *indefinite* would certainly be more appropriate here, but the word *particular*, with this special meaning, is the one which has been used traditionally.

With this two-fold distinction of quality and quantity we get four different kinds of propositions: *universal affirmative, universal negative, particular affirmative* and *particular negative.* For the affirmative propositions the letters A and I are used as symbols, A standing for the universal affirmative and I for the particular affirmative. E stands for the universal negative and O for the particular negative. (These letters are from the Latin *Affirmo* and *Nego.*)

- Propositions
  - Quality
    - Affirmative
    - Negative
  - Quantity
    - Universal
    - Particular

- Universal
  - Affirmative A
  - Negative E
- Particular
  - Affirmative I
  - Negative O

Propositions and Terms. The Relation of Subject to Predicate.—The question of the relation of propositions and terms is one that naturally arises here. A proposition obviously contains terms. Ordinarily it is said that a proposition is made up of two terms and a copula. One of these terms is the subject and the other is the predicate. The copula is that which connects subject and predicate; it is always some part of the verb *to be.* Some propositions do not fall naturally into this form: for example, "The earth moves." This can, however, be expressed in this form: "The earth is a body which moves."

This form, subject-copula-predicate, is called the "logical form" of the proposition. It often seems artificial, but for certain purposes it is convenient to employ it, and the attempt to restate propositions in this form is an excellent way of finding out just what the proposition means.

The subject of a proposition stands for that about which something is said.[1]

The predicate is that which is asserted of the subject. The copula is that which connects the two terms in a proposition; but the nature of that connection is not always the same. In the propositions, "Aristotle was the greatest pupil of Plato," "Aristotle was wise," "Aristotle was traveling in Asia Minor," and "Aristotle was a philosopher," the copula has, in each case, a different meaning. In the first, the relation is that of

[1] A distinction may be made between the grammatical and the logical subjects. The grammatical subject is the subject of the proposition; it is, as we have seen, a term. The logical subject has been variously defined. The definition of the logical subject as the subject of the *thought* seem, on the whole, to be the best. (See for discussion, Joseph, *Introduction to Logic.*) The logical subject is that about which the judgment is made. For example, in the proposition, "Acid turns blue litmus paper red," the grammatical subject is, of course, the word "acid." The grammatical predicate is that which stands for what is asserted about the subject; in this case, the words "turns blue litmus paper red." Changing the proposition into the form of subject-copula-predicate, it would read "acid is that which turns blue litmus paper red," and the complete predicate would be the words *following* the copula. Now the form of the proposition may not indicate the real logical subject. If the statement just given were the answer to the question, "What can you say

identity; in the second, that of subject and attribute; in the third, that of agent and action; in the fourth, that of inclusion of an individual in a class. Logicians have usually taken the last of these as the typical relation; the others can be transformed with more or less success into it. The proposition, "Aristotle was wise," can be put in the form, "Aristotle was a wise man"; and "Aristotle was traveling," etc., can be expressed in the form, "Aristotle was a man who was traveling," etc. These forms are sometimes criticised as not expressing the exact shade of meaning contained in the other forms; but the difference is usually not serious, and the performance of certain logical operations is much facilitated by so expressing the judgment as to indicate the inclusion of an individual, or a class, in another class.

In the negative proposition the relation will, of course, be that of exclusion. "Minors are not voters" indicates the exclusion of the first class from the second.

about *acid,* the grammatical and logical subjects would correspond; but if the question were, "What is the effect of acid on litmus paper?" the logical subject (*i. e.,* the thing about which the *judgment* is made) would be that which is expressed by the grammatical predicate of the proposition. The form of the sentence could be so changed as to make the grammatical or verbal subject correspond to the logical subject; in a great many cases they do not so correspond. Ordinarily the logical subject can be determined only by the context, though sometimes it can be indicated by emphasis on certain words. For example, "Acid turns blue litmus paper *red*" would imply, as the subject, what is expressed by the words, "the color to which blue litmus is turned by acid." Unless otherwise specified the term *subject* will be understood to mean *grammatical* subject; and *predicate* will mean the term that is joined to the subject by the copula. In the treatment of isolated propositions there is no occasion for the distinction. It is sometimes said that reality as a whole is the logical subject of every judgment. It might better be called the *ultimate* or *metaphysical* subject.

The Distribution of Terms in a Proposition.—There are degrees of inclusion and exclusion. In the illustration just given the whole of the class minors is excluded from the whole of the class voters. In the proposition, "Some men are not good citizens," only some men or a part of the class men is excluded from the class good citizens, but the whole of the class good citizens is excluded from that part of the class men which is included in the subject. In the proposition, "Some men are healthy animals," a part of the class men is included in the class healthy animals, and consequently a part of the class healthy animals may be included in the class men. Again, in the proposition, "All men are bipeds," the whole of the class men is included in the class bipeds, but so far as this proposition informs us, only a part of the class bipeds can be included in the class men. Whenever, in a proposition, a term is used to indicate the whole of the class for which it stands, it is said to be *distributed;* when it covers only a part of the class, it is *undistributed. The subject of a universal proposition is always distributed,* because, by definition, a universal proposition is one which asserts something about the whole of its subject. It will be seen in the examples given above that both the *negative propositions distribute their predicates.* That is always the case with negative propositions. They always indicate the entire exclusion of the predicate from the subject. The proposition *A*, being *universal* and *affirmative*, will *distribute* its *subject* but *not its predicate; I*, being *particular* (indefinite) and *affirmative,* will distribute *neither;* the *particular negative, O,* will distribute the *predicate*, but *not* the *subject*, while the *universal negative*, *E*, will distribute *both subject and predicate*.

Euler's Method.—Euler, a Swiss mathematician of the Eighteenth Century, devised the following method of representing the relation of subject and predicate and the distribution of each. Let each term be represented by a circle; then the E proposition will be represented as follows, S standing for the subject and P for the predicate: ⓈⓅ. S and P are shown to be entirely excluded from each other; each is distributed. The A proposition would be represented in this way:

S is seen to be entirely included in P, while, so far as we know, only a part of P falls within S; S is distributed, P is not. In the I proposition the circles would overlap. Each would be partially in-

cluded within the other; that is, neither would be distributed. Whether either extended further would be left undetermined; there are four possibilities, in each of which at least *some* S is *some* P. In

the O proposition, *part of S* would be excluded from P; *the rest would* be left *undetermined;* while all of P would be outside the specified part of S; S is not distributed, P is.

The distribution or non-distribution of terms in the various propositions may be represented by the following symbols: "-" indicating an affirmative proposition, "x" a negative one, and a circle about a term the fact that it is distributed.[2]

A, Ⓢ-P; E, Ⓢ x Ⓟ; I, S-P; O, S x Ⓟ

[2] The last two of these symbols are adapted from Hyslop, *Elements of Logic*.

Ambiguous Propositions.—There are several kinds of ambiguous propositions. In the first place the arrangement of the words and phrases may be such as to admit of two interpretations. Familiar examples are the prophecy in Shakespeare's *Henry VI*, "The Duke yet lives that Henry shall depose," and the response of the oracle, "Pyrrhus, I say, the Romans shall subdue." The expression, "Twice two and three," is ambiguous for the same reason, and so is the statement, "He went away and returned yesterday." In the two last, punctuation would, of course, remove the ambiguity. Propositions, like "He jests at scars who never felt a wound," will sometimes mislead a careless reader. Care in the construction of a proposition will obviate such difficulties; where such sentences are found, only the context can make clear what the meaning is. Wrong conclusions in such cases are said to result from committing the *Fallacy of Amphiboly* or *Amphibology*.

Certain other cases of ambiguity might be brought under this heading. One of these is found in the use of "all . . . not." In the statement, "All these men are not swift-footed," it might be thought that the meaning was, "None of these men is swift-footed"; that is, that the subject, *these men*, was distributed, and that the proposition was an E proposition. It is usually interpreted as meaning "*some* are not swift-footed," not "*all* are." It is an E proposition in form, but an O proposition in meaning. Again, it might seem to imply that "some *are* swift-footed"; but this last implication is not to be trusted, for we could make the original statement if we knew that some of these were not swift without knowing anything about the rest. The word some, as already noted, is indefinite; in an affirmative proposition, such as "Some are going," it seems to imply the corresponding negative, "Some are not going," and vice versa; but these implications count for nothing if not confirmed in some other way. In interpreting a proposition the only safe rule is to *include in its meaning only what it must mean, not what it may mean.*

In still other cases it is not so much the arrangement as it is the *character* of the terms that occasions difficulty. Propositions which are introduced by the word *few* are ambiguous. "Few are completely masters of themselves" really means that most are not masters of themselves, or that not many are. It is an O proposition in meaning, though like an I proposition in form. It may also seem to *suggest* the corresponding I proposition, "Some are masters," etc. The importance of making clear the negative force of such a proposition may be illustrated thus: suppose we have also the statement, "All who are masters of themselves are mature individuals"; it might seem that we could conclude that few are mature individuals. If the proposition "Few," etc., be put in the negative form given above there will be no temptation to draw the erroneous conclusion. Thus in, "Most men are not masters, etc.; those who are, are mature," etc.; the conclusion, "Most men are not mature," does not even seem to follow. Professor Hyslop, in his *Elements of Logic*, calls propositions of this sort *partitive propositions*, because they "express a part of a whole of which the implied proposition is a complementary part."

Another sort, similar in certain respects to these, is found in the *exclusive propositions;* they are such as have their application determined by such expressions as *only, alone, none but,* and the like. "None but native-born citizens are eligible to the presidency," "Only students will be admitted," etc., are exclusive propositions. These statements do not mean that all native-born citizens are eligible nor that all students will be admitted. They are not universal propositions; they do not distribute their subjects. They are equivalent to the propositions, "Those who are not native-born are not eligible," "Those who are not students will not be admitted," which are the complementary opposites of the original propositions, and are in this case E propositions. As the original propositions stand they really limit the application of their predicates, *i. e.*, they include the whole of the predicate in the subject. Thus they distribute the predicate, in spite of the fact that they are affirmative propositions. They are, therefore, exceptions to the rule for affirmative propositions (p. 71). Another way of restating exclusive propositions is to *convert* them, making the converse a universal proposition. Thus: "All persons eligible to the presidency are native-born citizens"; "All who are to be admitted are college students."

One other sort of proposition of this general class may be mentioned, the *exceptive proposition:* it is introduced by such words as "All except," "all but," etc. For example: "All but the best will be excluded." In addition to the positive statement, a corresponding negative is suggested, namely, "The best will not be"; though this last is not certainly true. It is well to restate such propositions, eliminating the exceptive particle. Thus: "All those who are not the best," etc.

Figurative statements are peculiarly liable to misinterpretation; Hyperbole and metaphor, symbolical and allegorical statements, may all be mistaken for literal statements, or if recognized as figurative, they may be wrongly interpreted on account of the inherent vagueness of most figurative expressions. Fallacies arising from this cause are known as *Fallacies of Figure of Speech.*[3]

Another source of ambiguity and misinterpretation in propositions is to be found in *misplaced emphasis.* Wrong emphasis gives rise to what is known as the *Fallacy of Accent.* To quote from Jevons: "It is curious to observe how many and various may be the meanings attributable to the same sentence according as emphasis is thrown on one word or another. Thus the sentence, 'The study of Logic is not supposed to communicate the knowledge of many useful facts,' may be made to imply that the study of Logic *does* communicate such a knowledge although it is not supposed to do so; or that it communicates a knowledge of a *few* useful facts; or that it communicates a knowledge of many *useless* facts. . . . Jeremy Bentham was so much afraid of being misled by this fallacy of accent that he employed a person to read to him, as I have heard, who had a peculiarly monotonous manner of reading." To introduce italics into a quotation, with no mention of the fact that they did not

3 It has been said that some of Locke's erroneous conclusions, in his Essay on the Human Understanding, resulted from his own use of "the sheet of white paper" as a figure representing the mind before experience has begun.

occur in the original, is usually to misrepresent the meaning of the original. De Morgan and others have pointed out that taking words or passages out of their context may have the same consequences. Isolated texts from sacred writings are often misused in this way; *e. g.*, "Eat, drink and be merry, for to-morrow ye die," "Take no thought for the morrow," etc.

Quoting an argument which an author has presented only in order to refute it, without mention of his purpose, is another case of the same sort.

EXERCISES

1. In each of the following propositions give (a) the complete subject and (b) the complete predicate; (c) restate each in its logical form; (d) give its quantity and quality and the letter which symbolizes it. And state whether (e) the subject and (f) the predicate are distributed.

(1) He laughs best who laughs last.
(2) Few are able to endure such hardships.
(3) Not all who are called are chosen.
(4) Nothing of worth is without honor.
(5) Only genius could have accomplished it.
(6) He little knows you, who can speak of you in such terms.
(7) Every bit of success makes further success easier.
(8) Like cures like.
(9) There is nothing either good or bad but thinking makes it so.
(10) It is the first step that costs.
(11) Everything has its limit.
(12) We demand non-partisan judges.
(13) His lack of enterprise cost him his position.
(14) The plowman homeward plods his weary way.
(15) Contentment is better than riches.
(16) Every deed returns upon the doer.
(17) All's fair in war.

(18) Many a morning on the moorland have we heard the copses ring.
(19) Perfection is beyond the reach of man.
(20) My mind to me a kingdom is.
(21) No admission except to ticket holders.
(22) Every one has the defects of his qualities.
(23) Socrates taught that no man would knowingly do wrong.
(24) And silence, like a poultice, comes to heal the wounds of sound.
(25) All the world admires heroism.
(26) It rains.

CHAPTER VI

INDUCTION

Generalization and What it Involves—In the two last chapters we have been studying the use of language, the most important instrument of thought. We now return to the consideration of the further processes mentioned in our preliminary survey of scientific method. (Chap. I.) Observation and classification have been discussed already; it remains to examine the way in which laws [1] can be discovered after considerable body of facts has been observed and classified. We have seen that a law is a statement of the way in which facts of a certain kind behave, how they are related to other facts, what are the universal relations in which they stand. Our first question is: How are these laws suggested? What is the source of these general statements of relationship? And our second question is: How is a supposed law to be established or verified?

In answer to the first question, it may be said that a general statement is usually arrived at by generalizing some observed relationship. If we have observed one or more instances in which a cold winter has been followed by a hot summer, we may generalize the connection and assert that a cold winter is always followed by a hot summer, that there is an invariable and inevitable

[1] The previous chapter has dealt, in part, with universal propositions; the present one is a discussion of the way in which such propositions are established.

connection between them. And similarly in any other case: A has been followed by B and we conclude that *every A is followed by B*, that every A has its B. A single instance may be enough to suggest a generalization. A generalization is a universal assertion, not a mere attitude of expectation. The lower animals exhibit a tendency to *expect* a given thing when another, which has occurred along with this, reappears; when an animal hears a certain call he may expect food, because in the past the two have been connected; when he sees a blow descending he may expect pain, and so on. But to expect a thing on the recurrence of another formerly connected with it, is not the same as to infer a universal connection. When I perceive A, I may remember and expect B, without ever having thought of a universal relation between the two, without asserting that B always follows A. There is no proof that an animal can generalize, that he can think to himself: "A call of a certain sort is followed by food," "A blow causes pain," and so on. He hears the call and expects food, but he does not generalize the connection. To do the latter would be difficult if not impossible, without language. This power to generalize, to use general and abstract ideas, is usually regarded as one of the most important differences between human and animal intelligence.

Without generalization, our knowledge would be confined to individual facts or to groups of these. We have seen that knowledge is not completed by the mere accumulation of observations and the classification of what has been observed. The aim of science is usually said to be *the discovery of laws*. Now a law, as already remarked, is the statement of the way in which phenomena behave, or the way in which they are inevitably related to other phenomena.

This tendency to generalize is, then, a pre-condition of all but the most primitive kind of knowledge. But, of course, our generalization may not turn out to be a law. *A law states a universal connection which actually holds true*, whereas our hasty generalization may be entirely unsound. A generalization arrived at in the way described above is an *inductive inference*. An inductive inference is a judgment about a whole class of facts based upon the observation of individual cases. It is a universal conclusion based upon one or more particular instances. Obviously, *an inductive inference must be tested or verified;* but before proceeding to the discussion of verification it will be well to mention certain other terms which are frequently employed in this connection.

Causal Connection.—The terms, *cause, causal connection, causal law*, occur constantly in this part of scientific method. What is a cause? The term implies a connection of some sort between phenomena; but of what sort? In ordinary usage it probably means most frequently something which produces or brings about something else. It has been objected that we can never observe one thing producing another; that we can at most observe that one thing is followed by another, and perhaps find reason for believing that it will always have such a connection; and that to say that A *produces* B, is to raise a metaphysical question with which science and everyday thinking are not concerned. But if we give up this way of conceiving cause, what can we put in its place? Is it sufficient to say that cause means simply invariable succession? No, for the succession of day and night is an invariable succession. The notion of cause implies that the relation of cause and effect not only *is* invariable, but also that it *must* be so; that there is an *unconditional* or necessary connection between the two; that if the first does not happen, the second cannot. In the field of physical phenomena, it is held also that the amount of energy in the cause is exactly equal to the amount of energy in the sum-total of its effects; in other words, that no energy is either lost or created. This is known as the Law of the Conservation of Energy. Whether it applies where mental phenomena are concerned has been questioned. However this may be, cause always means unconditional connection.[2] Two things stand in a relation of causal connection when they are so related that one is the unconditional accompaniment of the other; the cause usually occurs or begins before the effect, but there are cases in which both seem to begin together. Heat is a cause of expansion, but a body does not first become hot and then expand; the two phenomena occur simultaneously. A causal law is a statement, in general terms, of a causal connection. Thus: "Heat causes expansion."

The sort of generalization which is most frequently of interest and importance is one which asserts a connection of this sort, although this is not the only sort which may be investigated. For instance, the universal

2 Of course, the fact that a connection is unconditional cannot be observed. The reasons for asserting that it is unconditional will appear presently.

concurrence of two properties in a given substance may be a matter of importance, but their connection would not be called a causal connection. The specific gravity and the atomic weight of carbon would be a case in point. The co-existence of gravity and inertia is another example mentioned by Bain; the sciences furnish innumerable instances of a like sort. Bain remarks that "there are very few general laws of pure co-existence; causation is singular in providing a comprehensive uniformity that may be appealed to deductively for all cases. The uniformities of co-existence (independent of causation) can be proved only piecemeal; each stands on its own evidence of observation in detail; no one assists us to prove another." The causal law is the one to which we shall give most attention.

Testing Inductive Inferences.—We return to the question, "How can we prove an inductive inference to be true? How can we show that it is a law?" There are several things which would show that it was *not* true; if we found that there were *facts* which were *inconsistent with it*, or if it were found to be *inconsistent with itself*, or if it proved to be *in disagreement with any established law*, it could be rejected at once. But suppose that none of these things were found; should the inference be accepted as true? Not necessarily; it might be that our observation or our reflection on the case had been *insufficient to show us exceptions* or inconsistencies if such did exist. If we have inferred a universal connection we are very likely to overlook exceptions or to forget those which we have observed. A good many people still believe that Friday is an unlucky day and that the number 13 brings misfortune.

But even if no exceptions have occurred and if the inference is not inconsistent with known laws, *how can we be assured that exceptions may not occur in the future, or that further reflection might not discover fatal inconsistencies?* A large number of favorable cases is not alone sufficient to give this assurance. The example of the succession of day and night illustrates that. Millions of cases do not prove inevitable connection. On the other hand, a single experiment made by some scientist in his laboratory may be sufficient to establish some very important law. "Why," asks Mill, "is a single instance in some cases sufficient for a complete induction, while in others myriads of concurring instances, without a single exception known or presumed, go such a very little way towards establishing an universal proposition? Whoever can answer this question knows more of the philosophy of logic than the wisest of the ancients, and has solved the problem of Induction." Mill himself aided very materially in the formulation of the conditions under which we do regard our inductive inference as established, and the inductive methods presently to be discussed are usually called "Mill's methods." Whether or not he has solved the whole problem of induction is another question and one with which we shall not at present concern ourselves.

Complete Enumeration.—There are cases in which the establishment of universal conclusion might seem to be comparatively easy. It is sometimes possible that all the cases of a given sort may have been observed. For example, observation has shown that Mercury revolves about the sun; that Venus does also; and likewise of the Earth, Mars, Jupiter and each of the other planets. We can say then with perfect safety, that all the planets[3] revolve about the sun. The universal statement is warranted because each of the instances which it covers has been observed. We are saying no more in the conclusion than we had already said in the several statements on which the conclusion is based. The universal is, in fact, simply a summary way of expressing what had already been said. It is merely a "telescoping" of the other statements, as it were. This act of basing a general statement on a complete enumeration of the particular cases which it covers has been called *Perfect Induction.* It was so called because the conclusion is one which possesses complete certainty, whereas most inductive inferences are more or less uncertain. It might seem then that this was the solution of the problem raised above; you are sure of your universal if you have seen all the particulars which it covers. But how can we be sure that we have counted all the particulars? The field of observation may be so small and so easily explored that every existing case may be observed. But even if all existing cases have been observed, how can we be sure that others may not arise, and that they may not differ from those we have observed? We may have such knowledge about a class of objects as will enable us to say that if any other members of the class should come into existence they would be like those already known. We may know that the sum of the angles of every plane triangle which may ever exist will be equal to two right angles, not because we have counted cases, but because we know that this necessarily follows from the properties essen-

[3] Leaving the asteroids out of consideration.

tial to all triangles. However numerous the class which has been completely observed, the knowledge that each of the observed members stands in certain relations does not by itself assure us that other conceivable members of the class would be like them in this respect. Complete enumeration is useful as an abbreviated way of stating certain kinds of information, but it throws no light on the methods of discovering unconditional connections.[4]

The judgments which result from the complete enumeration of cases have been called, by some writers, Enumerative Judgments and by others, Collective Judgments.

How Generalizations can be Verified—It appears then that enumeration of all the existing members of a class does not enable us to establish laws. Anything short of that might seem to leave us still farther from that goal. And it is of course true, as appeared on page 84, that an incomplete enumeration of instances furnishes no verification. Then if verification is possible at all it can not rest on mere enumeration, or counting of cases. Suppose that the observation of one

[4] It may be well to note one case in which a statement in the universal form must be distinguished from a law. As an example, we may take, "Every three-sided figure is a triangle." This is not an inductive inference; it is not based upon the observation of individual instances at all. It is true in all cases, but it is true because we have previously said, "If any figure has three sides we will call it a triangle." In other words, it is true *by definition*. It is like an inductive generalization in applying to all possible cases, past, present and to come, real and imaginary, etc., but it is not based upon the observation of individual facts. Other judgments and operations, which must be distinguished from those which are present in induction properly so called, will be discussed in a later chapter; and still others may be found by referring to Mill's *Logic*, Book III, chapter ii.

or more instances in which B has followed A has suggested to us the inference that A and B are causally related. Let us ask ourselves what consequences would follow upon the truth of this inference. In the first place we could conclude that if B were present in any case A must have been present also; and, again, if A were absent in any case, its supposed effect B must have been absent also; or if either A or B varied in amount or degree the other should show a corresponding variation. All these things should be true of phenomena which are causally related. Phenomena which failed to satisfy such conditions could not be unconditionally connected. Suppose we had inferred that absence of oxygen would cause death. If that is true, an animal immersed in nitrogen should die. If experiment showed that an animal could live under such conditions, our inference would, of course, be disproved; but suppose the animal did die, would the inference be proved? Not necessarily. Perhaps the nitrogen acted as a poison or perhaps the death of the animal was due to rough handling, etc. Our inductive inference would be completely verified only if we could show that death could not have been due to anything except the absence of oxygen. If we could be sure that all the circumstances which were present before the experiment remained precisely the same with the one exception that oxygen was present in the first case and absent in the second, then we should have shown a necessary connection between the absence of oxygen and the occurrence of death. Nothing else could have been the cause because all were present when death did not occur. If a second circumstance were present when the phenomenon occurred and absent when it did not occur, it would dispute with the first the right to be called the cause and no final conclusion would be possible. When all other possibilities can be excluded, the one which remains is the cause. When no other inference is consistent with the facts, the one which is consistent must be accepted as true.

We can say, then, that an inductive inference is completely verified when we have found facts which are consistent with its truth and inconsistent with any possible rival inference; or more briefly, when it fits the facts and no alternative inference does. We establish one inference by eliminating all others. We reason that the phenomenon under investigation has some cause; this other phenomenon, A, may be the cause; it fulfills the requirements and no other does; therefore, this one is and must be the cause in question. There are several ways of selecting or grouping instances so as to show that some one factor alone satisfies the requirements. These are known as the INDUCTIVE METHODS.

Observation and Analysis are Presupposed.—One thing should not be forgotten. It is that the application of such principles as these presupposes very careful observation; if we are to be certain that no other circumstance is present when a given phenomenon is present or absent when it is absent, we must have observed all the other circumstances. In ordinary observation we note only a few of the circumstances; if we are untrained observers it may be impossible for us to observe more than a few. It is quite impossible for a child to observe in a flower all that a trained botanist can observe there. Accurate observation presupposes analysis, *i. e.*, breaking up the total complex phenomenon into its element. The beginner in any science is unable to handle the facts properly because he is unable to analyse them; he sees only their most obvious characteristics.

Postponing Inference till Test Conditions are Present—Before we begin the more detailed examination of the methods of verifying an inductive inference, there is one more statement to be made, namely this: instead of making a generalization and then searching for means of verifying it, *we may refrain from drawing any inference until we have before us a group of facts which will make it possible to draw a correct inference.* In other words, we may *make our inference under test conditions.* Suppose, for example, that we are trying to discover the cause of eclipses. Before making any theories on the subject we might observe a number of cases. If we found that whenever an eclipse occurred there was an opaque body between us and the source of light and that at other times everything was the same except that there was no body in that position, we should infer at once that the presence of the opaque body in that position was the cause of the eclipse. And so in any other case we might form no theory until we had facts which would make it possible to form a correct one.

"If a chemist discovers a new element, he will proceed to try a variety of experiments in order to determine the proportions in which it will combine with other elements as well as to discover the various properties of such combinations. Supposing such experiments to have been properly conducted, the inductions at which he arrives will be perfectly valid, though he may have formed no previous theories as to the results of his researches. Occasionally, too, an induction will not be the result of any definite course of investigation, but will be obtruded on our notice." [5] But such cases are rare, and ordinarily we have some theory before we have the facts which will verify it.

It is often better to draw an inference early in the investigation; the reasons for this will be discussed in a later chapter.[6] In the meantime it should be remembered that the Inductive Methods which are now to be discussed may be used either to test an inference already made or to furnish a basis for drawing a correct inference if none has previously been drawn.

The Inductive Methods.—I. AGREEMENT. Suppose we find that A was followed by B in a number of instances, but that the attendant circumstances varied greatly. Suppose, for example, that three or four individuals, of different races, different habits of life, and otherwise as different as possible, were all bitten by a certain kind of mosquito and that each developed yellow fever: would not such a set of cases give some warrant to the inference that the bite of the mosquito was causally connected with the development of the fever? The fact that these individuals were different in all other respects would seem to exclude the possibility that anything else could have been the cause.

Or suppose again that we find dew deposited on two or more objects which differ in position, chemical composition, character of surface, and, in short, all respects except that both are cooler than the surrounding atmo-

5 Fowler, *Inductive Logic*, p. 11.

6 Part III, chapter ii, Hypothesis.

sphere; we should have good grounds for believing that this last characteristic was causally related to the deposition of dew.

Or, once more, if a number of persons who recovered from a given disease were similar only in having used a certain drug, the inference would be that the drug was causally related to the cure.

In each of these examples we have a set of instances in which a given phenomenon is present, an attack of fever, deposition of dew, recovery from illness; nothing else is present in all cases except one other phenomenon, being bitten by the mosquito, being cooler than the surrounding air, having used a certain drug. Our inference is that the phenomena which are constantly present together are causally related. If A is causally, *i. e.*, unconditionally, related to something else, that thing must be present when A is present. As only one other phenomenon is present in all cases, that alone among all those at any time present can be causally related to the first. This method of isolating the phenomena which are so related is known as the Method of Agreement. Mill's statement of the *Canon of Agreement* is as follows: "If two or more instances of the phenomenon under investigation" [fever, deposition of dew, etc.], "have only one circumstance in common" [being bitten by a mosquito, being cooler than the surrounding air, etc.], "the circumstances in which alone all the instances agree is the cause (or effect) of the given phenomenon." His statement of the axiom of this method is: "*Whatever circumstance can be excluded without prejudice to the phenomenon, or can be absent notwithstanding its presence, is not connected with it in the way of causation.*" The only circumstance which is common to a number of instances in which a given phenomenon is present, is causally related to it, because all the rest are excluded by the fact that they are separable from it.

In this method, as in the others to be discussed, the point of first importance is that we are selecting instances of the occurrence of a phenomenon, and selecting them in such a way as to identify the circumstance or circumstances causally related to the phenomenon. We might re-word the main points as follows: select instances in which the phenomenon under investigation is present, but which are as different as possible in all other respects; if there is one circumstance and one only which is always present when the phenomenon under investigation is present, that circumstance is causally related to the phenomenon.

Difficulties in Using this Method.—An ideal case might be represented symbolically in this way: let the phenomenon under investigation be represented by *x;* and let the accompanying circumstances, in the several instances we have selected be represented by *abcde, afghi, ajklm,* respectively; or the phenomenon and its accompanying circumstances by:

abcdex
afghix
ajklmx.

The only circumstances common to all the instances is *a.* Therefore *a* is causally related to *x.* No actual case would be quite so simple; any phenomenon has among its accompanying circumstances everything that is happening in the universe at the time of its occurrence. Most of these circumstances can, of course, be eliminated as irrelevant; still it is easily possible to overlook something that is relevant.

1. Circumstances which might seem to have no connection with the phenomenon may be causally related to it. For example, it might be supposed that the number of sun-spots had no relation to financial conditions; yet it has been shown that the periods when sun-spots are most numerous have the same frequency as the periods when panics occur; and it has been suggested that sun-spots influence climatic conditions, that these in turn, by influencing crops, and so on, do affect financial conditions. Whether there is any truth in this or not, it will remind us that it is not easy to determine just what circumstances are relevant.

2. Another difficulty arises from the fact that analysis is never complete; not all the elements are singled out, and some of those which have been overlooked may be all-important. For example: it had been noticed that persons who had been much out of doors at night were more likely than others to be attacked by malaria; it was inferred that "night-air" was a cause of malaria, and consequently people tried to exclude it from their houses. Later it was shown that the attacks of mosquitoes were the causes. Mosquitoes are more active at night, but instead of noting this, the more obvious fact that "night-air" is damp, and so on, was selected as the important one. When it was found that the bite

of the mosquito was the circumstance always present, while the time of the attack made no difference, the older theory was overthrown.

In symbolizing an actual case we should need something to represent the circumstances which were disregarded or overlooked. We might use the symbol *X*; the accompanying circumstances would then be represented by *abcde*. .*X*, *afghi*. .*X*, etc. This would indicate that there is a margin of uncertainty in such cases.

3. There is one other difficulty in the application of the method of Agreement. It may be illustrated in this way: suppose a man should drink coffee with his luncheon on one day and afterwards smoke a strong cigar; suppose on the following day, with a different bill of fare, he should drink tea and smoke a cigar; on both days he has a headache in the afternoon. The application of the method of Agreement would lead him to believe that the cigar was the cause of the headache, whereas the cause may have been the coffee in one instance and the tea in the other. This illustrates what is known as the *Plurality of Causes*. A given phenomenon may have one cause in one instance and another cause in a second instance. It is sometimes said that if our analysis were complete we should find that a given phenomenon always had the same cause; that in the instances just mentioned, the cause of the headache was something common to tea and coffee; or that the headache caused by coffee differs from that caused by tea and that two things which are different never can produce the same effect. Perhaps that is true, but the fact still remains that, in practice, effects which are so simi-

lar as to be indistinguishable may be produced by causes that, for ordinary observation, are very different. This makes a very serious limitation to the application of the method of Agreement. The method is still valuable as suggesting causal relations, though imperfect as a means of proof. May it not be possible to select instances on some other principle in such a way as to obviate some of these difficulties?

II. The Method of Difference.—Take another concrete case: suppose two individuals as similar as possible in all respects, race, family, occupation, manner of living, state of health and so on; one of these is bitten by mosquitos of a certain sort and the other protects himself against their attacks; the first contracts the fever, the second escapes it. We should regard such a group of facts as warranting the conclusion that the bite of the mosquito and the contraction of yellow fever were causally related.

Or, if two objects of similar chemical construction, character of surface, location, etc., differed only in that one was for some reason cooler than the surrounding air, while the other was not, and if dew were deposited on the first and not on the second, we should conclude that the cooler temperature of the one was causally related to the deposition of dew upon it.

Cases like these illustrate the Method of Difference. Mill's statment of the *Canon of this method* is:

"If an instance in which the phenomenon under investigation occurs, and an instance in which it does not occur, have every circumstance in common save one, that one occurring only in the former; the circumstance in which alone the two instances differ is the effect or the cause, or an indispensable part of the cause, of the phenomenon." Its axioms, in the words of the same writer, are: "*Whatever antecedent cannot be excluded without preventing the phenomenon, is the cause, or a condition of the phenomenon; whatever consequent can be excluded with no other difference in the antecedents than the exclusion of a particular one, is the effect of that one.*"

Relation of this to the First Method.—We quote from Mill again regarding the relation of this method to the other:

"Instead of comparing different instances of a phenomenon to discover in what they agree, this method compares an instance of its occurrence with an instance of its non-occurrence, to discover in what they differ. . . . Both are methods of *elimination*. . . . *The Method of Agreement stands on the ground that whatever can be eliminated is not connected with the phenomenon by any law. The Method of Difference has for its foundation, that whatever cannot be eliminated is connected with the phenomenon by a law.*"

Incomplete analysis of the circumstances attending the phenomenon may vitiate the inference in both methods: in the method of Agreement as already stated; in the method of Difference, by leading us to overlook points of difference in cases supposed to be alike except in the particulars specified in the Canon of this method.

Difficulties in Using the Method of Difference.—1. One danger in employing the method of Difference results from the possibility of the *Composition of Causes.* It often happens that a given phenomenon is the effect of the joint action of several causes. Heat, light, moisture, etc., are all causally related to the life of plants. If two plants were similarly situated as regards all but moisture, it would be incorrect to conclude that moisture was the sole cause of the life of the one because its absence was followed by the death of the other. Application of the method of Difference does show that the antecedent is causally related to the consequent, but not that it is the sole or adequate cause.

If we could supplement the method of Difference by the method of Agreement, if we could find a set of instances in which the supposed cause was alone common to the cases in which the phenomenon was present, then we could conclude that this supposed cause was adequate to produce the phenomenon.

2. Another source of difficulty closely connected with the one just discussed is to be found in the existence of *Counteracting Causes.* Even if a cause is adequate to the production of an effect under ordinary conditions, it may fail to produce it owing to the presence of some opposed tendency. The poison can be counteracted by an antidote; the tendency of the moon to fall to the earth is in part overcome by centrifugal force, and so on. The presence of a counteracting cause might lead us to overlook the real cause of the phenomenon. The cause is present without the effect; that is, without the usual effect. In such a case its effect is to be found in its modification of the counteracting cause. The tendency of the moon to fall does modify the effect which would otherwise be produced by its tendency to fly off at a tangent. We may symbolize one sort of case in which error might arise, as follows: Let *m* be the phenomenon

whose cause we are seeking and suppose that we have

abcd X
abcr X m

We should probably conclude that *r* was the cause of *m*, whereas it might well be that *a* was the real cause, but that it was counteracted in the first instance by *d*. It is no doubt true that in an ideal application of the methods there would be little difficulty; if we could get cases which differed in only one circumstance it would at any rate be easy to see that the absence of the effect *m* was connected with the presence of the circumstance *d;* we should still have to search for the cause of *m's* presence. But in actual cases the matter is not so simple; we cannot find ideal cases and the instances we select for the application of the method of Difference may differ in more than one respect, as in the case just discussed.

III. The Joint Method.—Nature presents very few instances in which the method of Difference can be directly applied, and even experiment fails to present ideal conditions. It seldom happens that the conditions stated in the Canon of Difference are realized. The same is true to a great extent with regard to the method of Agreement. Usually two or more cases in which a given phenomenon occurs are similar in more than one circumstance. In such cases it is sometimes possible to use a combination of the two methods. The following instance illustrates the use of The Joint Method of Agreement and Difference: a large number of cases of typhoid fever occurred at about the same time in a college community. It happened that all those who developed the disease ate at a certain few fraternity and boarding-house tables. The water supply was first investigated. It was found that all these places used water from the same source. But it was also true that the other houses were supplied from the same source, so this possible cause was eliminated. The fresh vegetables were supplied from various sources; some of the places in which the disease was developed used one source, others a different one; moreover, the places in which the disease was not developed were supplied from the same variety of sources. The other food supplies came from various places and the method of Agreement could not be applied so far as they were concerned, with one exception; it appeared that the milk supply was the same for all the places in which the fever was developed, whereas none of the places which escaped used milk from that source. The inference was that the milk contained the cause of the disease. Further, it was found that when milk from this source was no longer used, no new cases of the disease appeared. There were two sets of cases: in one the disease was developed, in the other it was not. Those in which it appeared were alike in several respects; the ages, habits and previous general health were similar in all; the water supply was the same and also the milk supply; it might be any one of these; the method of Agreement could not be successfully applied. The other set of cases, those in which the disease did not appear, were like the first in many respects, but there was no one of these which differed from any one of the others, in one respect only. There was one and one only circumstance in which all the members of the first group differed from all the members of the second, namely in the milk supply. All of one group agreed in having a given milk supply and developing typhoid fever; all of the other group agreed in using milk from another source and escaping the disease. Comparing group with group the method of Difference could be used. There was only one circumstance in which all the instances in which fever was developed differed from all of those in which it was not developed. Within each set of instances there is a partial application of the method of Agreement. One set agreed in having the disease and also in having another common circumstance; but more than one circumstance was common, so the application could not be complete; similarly the other set of instances agreed in the absence of the fever and in the absence of this circumstance; but they also agreed in lacking various other circumstances. However, they agreed in lacking only one which was present in all those of the other set, and that is the important point.

The two sets of instances might be symbolized thus; *p* representing the phenomenon under investigation:

abcdr X p	*bckr*
abefr X p	*eflr*
aefgr X p	*fgmr*
afghr X p	*ghnr*

Both *a* and *r* are common to all the instances in which *p* is present, but *r* is excluded by the fact that it is present in those in which *p* is absent.

Mill's statement of the *Canon of the Joint Method* reads: "If two or more instances in which the phenomenon occurs have only one circumstance in common, while two or more instances in which it does not occur have nothing in common save the absence of that circumstance, the circumstance in which alone the two sets of instances differ is the effect or the cause, or an indispensable part of the cause, of the phenomenon." This statement does not quite cover a case like that described above. The instances in which the phenomenon occurs may have more than one circumstance in common, provided that there is only one which is at the same time common to these and absent from those in which the phenomenon does not occur. We might restate it thus: "If two or more instances in which a phenomenon occurs have in common one circumstance which is at the same time the only circumstance present in these instances and absent from two or more instances in which the phenomenon does not occur, that circumstance is causally related to the phenomenon." This form of statement avoids the difficulty just mentioned and also another. It is practically impossible to find a set of instances which have nothing in common save the absence of one circumstance. In the example just given, the instances in which typhoid fever did not occur agreed in not being Esquimaux nor octogenarians nor coal-miners, and so on indefinitely. On the other hand it would have been easily possible to find a group of instances in which there would have been fewer circumstances absent from all. One might select a number of individuals from different races, of different ages, occupations, and so on. Fewer circumstances would be absent from all of these than from a homogeneous group of college students. But such instances might be entirely insignificant for the purpose of discovering the cause of the disease. Of course, if the group of negative instances included examples from all varieties of those who lacked the phenomenon in question, and we could discover the only circumstance lacking in these, and present in the cases where the phenomenon was present, a conclusion could be drawn; but, in the first place, it would be impossible to get such a group, for it would be infinite in extent; and, secondly, if the group could be had, the discovery of the only circumstance lacking from all of them would be an endless task. Most of such instances might at once be eliminated as irrelevant, though Mill's canon does not provide for that. It is important that the instances in which the phenomenon is absent should be similar to those in which it is present, for if there are many points of difference it will be difficult or impossible to select those which are causally related to the phenomenon.

IV. Concomitant Variations.—There are still other methods of discovering causal relations. Suppose a case in which such instances as are demanded for the application of any of the foregoing methods cannot be obtained; it may be possible to find instances in which the phenomenon occurs in varying degrees or in different quantities, while some other phenomenon varies concomitantly. "The effects of heat are known only through proportionate variation. We can not deprive a body of all its heat; the nature of the agency forbids us. But by making changes in the amount, we ascertain concomitant changes in the accompanying circumstances, and can so establish cause and effect. It is thus that we arrive at the law of the expansion of bodies by heat. In the same way we prove the equivalence of heat and mechanical force as a branch of the great law of Conservation of Persistence of Force."

"The proof of the First Law of Motion, as given by Newton, assumed the form of Concomitant Variations. On the earth, there is no instance of motion persisting indefinitely. In proportion, however, as the known obstructions to motion—friction and the resistance of the air—are abated, the motion of a body is prolonged. A wheel spinning in an exhausted receiver upon a smooth axle runs a very long time. In Borda's experiment with the pendulum, the swing was prolonged to more than thirty hours, by diminishing the friction and exhausting the air. Now, comparing the whole series of cases, from speedy exhaustion of movement to prolonged continuance, we find that there is a strict concomitance between the degree of obstruction and the arrest; we hence infer that if the obstruction were entirely absent, motion would be perpetual. The statistics of crime reveal causes by the method of Variations. When we find crimes diminishing according as labor is abundant, according as habits of sobriety have increased, according to the multiplication of the means of detection, or according to the system of punishments, we may presume a causal connection, in circumstances not admitting of the method of Difference."[7]

We may symbolize a set of instances to which this method is applicable in this way:

$$\begin{array}{llll} a & bcd & X & p \\ (2a) & bce & X & (2p) \\ (4a) & bef & X & (3p), \text{ etc.} \end{array}$$

The *Canon of the Method of Concomitant Variations*

[7] Bain, *Logic*, pp. 62-63.

is: "Whatever phenomenon varies in any manner whenever another phenomenon varies in some particular manner, is either a cause or an effect of that phenomenon, or is connected with it through some fact of causation."

V. The Method of Residues. This method is usually included with the others and completes the list. Its canon is: "Subduct from any phenomenon such part as is known by previous inductions to be the effect of certain antecedents, and the residue of the phenomenon is the effect of the remaining antecedents." Its principle, like that of the other methods, is that of exclusion. If we have a complex phenomenon or a group of phenomena represented by the letters *xyzlm*, and a group of antecedent circumstances represented by *abcdf*, and if we know that *a* causes *x* and *b* causes *y*, *c* causes *z* and *d* causes *l*, the conclusion will be that the remaining *m* is the effect of *f*. This method would be equally applicable in an instance in which the causes were present with the effects instead of being antecedent to them only. Thus, if we had a phenomenon *m* in a group of circumstances, *abcdfxyzlm*, and knew as before that *a* and *x*, *b* and *y*, *c* and *z*, and *d* and *l* were causally related, the connection of *f* and *m* would be evident. We must, of course, be careful to include all the relevant circumstances in the group.

If a phenomenon has occurred and all the known antecedents of this phenomenon are known not to contain its cause, the cause must be sought for in some phenomenon not yet discovered. There were certain perturbations in the movement of the planet Uranus, not accounted for by the attractive force of any known heavenly body; they must, then, be due to some body not yet discovered; this line of reasoning led to the discovery of the planet Neptune.

Again, the weight of atmospheric nitrogen was found to be greater than that of nitrogen produced chemically; further examination revealed the presence in the atmosphere of another element, argon.

Obviously the method of Residues can be applied only when we have fairly complete knowledge of the field of facts in which the phenomenon is found. We must know the causal relations of all the circumstances involved in the case except the phenomenon under investigation.

The following example, though not formally included under Mill's canon, employs the principle of Residues: If only four men were capable of doing a certain act and if we learned that one of these was temporarily unable to do it, through illness, and that the two others were a thousand miles away when the act was performed, the fourth must have committed the act.

If, *in any way*, we can assure ourselves that of all the possible causes of a phenomenon all but one are excluded, that one must be the cause. The several methods are ways of doing this.

EXERCISES

Examine the following arguments and criticise the reasoning as fully as possible; state the method used:

1. The newly discovered painting must be a Rubens; for the conception, the drawing, the tone and the tints are precisely those seen in the authentic works of that master. (Hyslop.)

2. In nine counties, in which the population is from 100 to 150 per square mile, the births to 100 marriages are 396; in sixteen counties, with a population of 150 to 200 per square mile, the births are 390 to 100 marriages. Therefore the number of births per marriage is inversely related to the density of population and contradicts Malthus's theory of the law of population. (Hyslop.)

3. The great famine in Ireland began in 1845 and increased until it reached a climax in 1848. During this time agrarian crime increased very rapidly until, in 1848, it was more than three times as great as in 1845. After this it decreased with the return of better crops until, in 1851, it was only 50 per cent. more than in 1845. It is evident from this that a close relation of cause and effect exists between famine and agrarian crime. (Hyslop.)

4. The influence of heat in changing the level of the ground upon which the temple of Jupiter Serapis stands might be inferred from several circumstances. In the first place, there are numerous hot springs in the vicinity, and when we reflect on the dates of the principal oscillations of level this conclusion is made much more probable. Thus, before the Christian era, when Vesuvius was regarded as a spent volcano, the ground on which the temple stood was several feet above water. But after the eruption of Vesuvius in 79 B. C., the temple was sinking. Subsequently Vesuvius became dormant and the foundations of the temple began to rise. Again Vesuvius became active, and has remained so ever since. During this time the temple has been subsiding again, so far as we know its history. (Hyslop.)

5. Take a bottle of charged water, slightly warmer than a given temperature registered by the thermopile, and mark the deflection it causes. Then cut the string which holds it and the cork will be driven out by the elastic force of the carbonic acid gas. The gas performs its work, and in so doing it consumes heat and the deflection of the thermopile shows that the bottle is cooler than before, heat having been lost in the process. (Hyslop.)

6. As an evidence of the extreme antiquity of highly civilized man, we have the following facts: On one of the remote islands of the Pacific—Easter Island—two thousand

miles from South America, two thousand miles from the Marquesas, and more than one thousand miles from the Gambier Islands, are found hundreds of gigantic stone images, now mostly in ruins. They are often forty feet high, while many seem to have been larger, the crowns of their heads, cut out of red stone, being sometimes ten feet in diameter, while even the head and neck of one is said to have been twenty feet high. The island containing these remarkable works has an area of about thirty square miles, and as the smallest image is about eight feet high, weighing four tons, and as the largest must weigh over a hundred tons or much more, their existence implies a large population, abundance of food, and an established government which so small an island could not supply. (Hyslop.)

7. We observe very frequently that very poor handwriting characterizes the manuscripts of able men, while the best handwriting is as frequent with those who do little mental work when compared with those whose penmanship is poor. We may, therefore, infer that poor penmanship is caused by the influence of severe mental occupation. (Hyslop.)

8. In the following instances crystallization takes place: the freezing of water; cooling and solidifying of molten metals and minerals; deposition of salts from solutions; volatilization of solutions; deposition of solids from the gaseous state, as iodine; pressure; slow internal change, as in rocks; the transformation of metals from the tough to the brittle condition, by hammering; vibration, and repeated heatings and coolings. We may then conclude that the cause of crystallization is the increased scope and operation of the molecular or solid-forming cohesion. (Bain.)

9. When the barometer was carried to the top of the Puy de Dome it was found that the mercury stood lower than before. It was inferred that the pressure of the air was the cause of the rise of mercury in the tube.

10. The chemical action between two substances is much greater when they are in a liquid than when they are in a gaseous state. We may conclude that there is an inverse relation between cohesion and chemical activity.

11. Goldscheider proved that muscular sensations play no considerable part in our consciousness of the movement

of our limbs, by having his arm suspended in a frame and moved by an attendant. Under these circumstances, where no work devolved on the muscles, he found that he could distinguish as small an angular movement of the arm as when he moved and supported it himself.

He also proved that the chief source of movement-consciousness is pressure-sensations from the inner surface of the joints, by having his arm held so that the joint surfaces are pressed more closely together, and finding that a smaller movement was now perceptible. (Creighton.)

12. "That the Tempest belongs to the latest period of Shakespeare's literary activity is shown, *inter alia,* by the absence of rhyme, the large number of 'run on' (unstopped) lines, the high proportion of weak and light endings, and the comparative rarity of puns in the low scenes." (Mellone.)

13. That the feeling of effort is largely, if not entirely, of peripheral origin, appears from such experiments as the following: Hold the finger as if to pull the trigger of a pistol. Think vigorously of bending the finger, but do not bend it. An unmistakable feeling of effort results. Repeat the experiment, and notice that the breath is involuntarily held, and that there are tensions in the other muscles. Repeat the experiment again, taking care to keep the breathing regular and the other muscles passive. Little or no feeling of effort will now accompany the imaginary bending of the finger. (Ferrier, quoted by Hibben.)

14. Sir Charles Lyell, by studying the fact that the river Ganges yearly conveys to the ocean as much earth as would form sixty of the great pyramids of Egypt, was enabled to infer that the ordinary slow causes now in operation upon the earth would account for the immense geological changes that have occurred, without having recourse to the less reasonable theory of sudden catastrophes. (Hibben.)

15. Count Rumford in 1798 proved that the common notion that heat was a substance was false, by boring a large piece of brass, under great pressure of the bore, whilst the brass was in a gallon of water; and at the end of two and one-half hours the water actually boiled. (Hibben.)

16. How would you set out to discover the causal rela-

tions of the following phenomena? Suggest instances and indicate the method to be used:

(1) Heat and expansion.
(2) Heat and friction.
(3) Mosquitos and malaria.
(4) The tubercle bacillus and consumption.
(5) Golden-rod and hay fever.
(6) A rainy spring and mosquitos.
(7) The presence of oxygen and the burning of a candle flame.
(8) Cocaine and the absence of pain.
(9) Moisture and vegetation.
(10) The gulf-stream and climate.
(11) The cause of the tides.
(12) The cause of the trade winds.
(13) The course of a glancing bullet.

17. Cite ten cases of the composition of causes.
18. Cite ten of the plurality of causes.
19. Cite ten cases of counteracting causes.
20. Bring in five cases illustrating each of the methods.

CHAPTER VII

VERIFICATION AND DEDUCTION

Verification and Deduction.—All these methods are means by which a sound inference may be drawn or an inference already drawn may be verified. They all involve finding certain facts which inevitably follow from the inference in question, and they are not conclusive if these facts can be shown to be consistent with any rival hypothesis.

There is another way of testing the truth of any inference; if we can show that the inference *follows from something already known* we shall establish the truth of the inference itself. Instead of searching for the consequences of the inferences and trying to determine their truth, we find a law of which *our inference is itself a necessary consequence.* Conversely if an inference is inconsistent with a known law it is necessarily false. In applying this it is necessary to remember that many supposed laws have proved to be false and that when an inference disagrees with a supposed law, it may be that the latter—or both—must be rejected. The fact that an inference is consistent with known laws does not prove its truth, but only its possible truth, for two rival hypotheses may be consistent with all the known facts and laws to which they are related. For proof, the connection must be closer than mere consistency. The inference must not only agree with the law, it must *follow* from it; in other words, the truth of the law must *insure* the truth of the inference.

An inference from a law or general principle to some consequence of the principle is a *deductive* inference. When we reason in this way we reason *deductively*, we *deduce* a conclusion, we employ *deduction*.

Systematic Knowledge.—When we show that an inductive inference is a reliable statement of the relation of certain phenomena to each other, or when we show that any inference whatever is a consequence of some general principle, we establish the fact that the inference with which we are dealing belongs to *a system of facts or truths*.[1] In a system all the parts and elements are so related that the truth of one part implies the truth of the rest; we cannot hold to one part and reject the rest without inconsistency and contradiction. A system may consist of comparatively few members and be comparatively simple, as in an isolated syllogism, or it may be very broad in its scope and its internal relations may be exceedingly complex. For example, a philosophical system attempts to state the laws which hold for all reality.

We shall begin our examination of systems with the syllogism. When we argue, to use the most ancient of illustrations, that "Socrates is mortal because all men are mortal and Socrates is a man," we are basing the truth of our conclusion upon a universal proposition, "All men are mortal," and the further proposition, "Socrates belongs to the class men."

Criticism of the Syllogism.—It might be urged, as an objection to the syllogism, that "it gives us no new

1 Professor Hibben in his *Logic, Deductive and Inductive*, makes much use of this conception in discussing the nature of deduction and induction.

information; if the conclusion is really contained in the major premise,[2] as it must be if the reasoning is to be valid, why go to the trouble of making a syllogism? We knew beforehand that all members of the class designated by the subject were included in that designated by the predicate, or possessed the quality, relation, or whatever it may be, for which the predicate stands; if we did not know that Socrates was mortal, how could we say that *all* men are mortal? Therefore, it is a matter of course that the subject of the conclusion, which is included in the subject of the major premise, will have that predicate. This objection would lead to the condemnation of such a science as geometry, for all its conclusions are contained in its postulates and axioms. Still we do get information by means of such processes.

We may know it to be a general law that all iron compounds have certain properties without knowing the chemical composition of a compound we have in our hands; as soon as we discover that it is a compound of iron, we can draw our conclusion. Of course, if our major premise were not a law, our conclusion would not be trustworthy. If the general statement about iron compounds were an unverified inductive inference, then we could not state it with certainty so long as we were not sure that the present compound, if it proves to be iron, would possess the given properties. If all inductive inferences were simply enumerative or collective judgments (page 86), if "perfect induction" were the ideal form of induction, then there would be ground for the objection we have mentioned. But if

[2] The universal proposition on which the reasoning is based, in this case, "All men are mortal," is called the *major premise*.

we may know that whenever a given phenomenon occurs, a certain circumstance must inevitably be present, or that any two properties are invariably connected, we have information which will apply to many cases of whose character we may yet be in ignorance. The syllogism is the typical form of reasoning. The quotation which follows is from Professor James's *Psychology;* it states the claim that reasoning is precisely that form of mental activity which does enable us to deal with new situations, with novel data.

"A thing inferred by reasoning need neither have been an habitual associate of the datum from which we infer it, nor need it be similar to it. It may be a thing entirely unknown to our previous experience, something which no simple association of concretes could ever have evoked. The great difference, in fact, between that simple kind of rational thinking which consists in the concrete objects of past experience merely suggesting each other and reasoning distinctly so-called is this: that whilst empirical thinking is only reproductive, reasoning is productive. An empirical, or 'rule-of-thumb,' thinker can deduce nothing from data with whose behaviour and associates in the concrete he is unfamiliar. But put a reasoner amongst a set of concrete objects, which he has neither seen nor heard of before, and with a little time, if he is a good reasoner, he will make such inferences from them as will quite atone for all his ignorance. Reasoning helps us out of unprecedented situations—situations for which all our common associative wisdom, all the 'education' which we share in common with the beasts, leaves us without resource." *Psychology, Briefer Course,* page 352.

As soon as we see that the present case belongs to a certain class or is of a certain type, the laws which are known to apply to that class or type may immediately be applied to it.

How Propositions are Related to each other.—The syllogism as illustrated above shows that a universal judgment may be made the basis for certain other statements. There are several kinds of syllogisms, but before discussing these it will be well to examine propositions generally with a view to discovering what relations different kinds of statements bear to each other and whether there may not be other ways than that illustrated in the syllogism, in which one statement may be made the basis for another. We have already discussed four kinds of propositions; those which are *universal* and *affirmative*, *universal* and *negative*, *particular* and *affirmative*, and *particular* and *negative*. It will be remembered that the symbols for these were A, E, I, and O respectively. Their relations to each other are best shown by means of what is known as the "Square of Opposition," a diagram which has remained practically unchanged since the time of Aristotle.

(All x is y) (No x is y)
A Contraries E

Subalterns Contra dictories Subalterns
Contra dictories

I Subcontraries O
(Some x is y) (Some x is not y)

Let us take as an illustration of the A proposition, "All men are rational." Its *contrary* will be "No men are rational."[4] What exactly are the relations between these two propositions? If A be true, it is obvious that E will be false, and if E be true A will be false. But if A be false, what about E? It may be true or false, for the falsity of A leaves those two possibilities; in other words, the truth or falsity of E is undetermined. Similarly of A, if E be false. If either be false, there is a middle ground; thus, it may be that some men are rational and some are not. *Two propositions are contrary when only one can be true and both can be false.*

"All men are rational" (A), and "Some men are not rational" (O), are *contradictory* propositions. If A be true, O will be false, and if A be false, O will be true; likewise if O be true, A will be false, and if O be false A will be true. There is no middle ground; there is no third possibility. Only one can be true, and only one can be false, or in other words, *both* cannot be true and *both* cannot be false. *Two propositions are contradictory when they are exact opposites; one must be true and the other must be false.*

Of sub-contrary propositions, both may be true, but only one can be false. The propositions, "Some men are rational" and "Some men are not rational," are in the relation of sub-contraries. If either be false the other must be true, and if one be true the other may be true also; both may be true, and one must be. The propositions I and O are consistent, whereas contraries and contradictories are inconsistent.

[4] Two propositions which are to stand in a relation of opposition to each other must have identical terms. This is true in the traditional treatment, but exceptions will be noted later.

In the case of *subalterns*, both propositions are of the same quality, but they differ in quantity. "All men are rational" and "Some men are rational" are subalterns. *If the universal be true the particular will of course be true also; but if the universal be false the other is left indeterminate;* it may be true or it may be false. On the other hand, *if the particular proposition be false, the universal will necessarily be false too;* if it is false that "Some men are rational," it cannot be true that "All men are"; but if the particular be true *it is by no means certain that the other will be.* Thus, if we know that some men are rational, that does not give us a right either to affirm or to deny that all men are rational; in other words, the truth of the universal is left indeterminate.

To summarize: contrary propositions are such that only one can be true, and both may be false.

Contradictory propositions are so related that one must be true and the other must be false.

Sub-contraries may both be true, but only one can be false.

Subalterns may both be true or both false. The truth of the universal assures the truth of the particular, but the falsity of the universal does not involve the falsity of the particular; the falsity of the particular involves the falsity of the universal, but the truth of the particular does not assure the truth of the universal.

Relations of Opposition among Propositions which have not Identical Terms.—These relations are most easily detected between propositions which have the same subject and the same predicate, but it is possible to find them in propositions which do not answer to this description. The propositions, " All men are rational " and " All men are idiots," are contraries. Both cannot be true, but both may be false.

Or again, " Socrates was the wisest of the Greeks " and " Aristotle was the wisest of the Greeks," are contrary propositions. Only one of them could be true, while both might be false.

This will be found to be the case with a great many inconsistent propositions. In fact all inconsistent propositions which are not contradictories are contraries. Many pairs of such inconsistent propositions would not fit into the square of opposition, for both might be affirmative or both might be negative. A and E propositions which have identical terms are contraries: we need not consider the *meaning* of the proposition to discover that, for it is evident from their *form*. In other cases we must take into account the *meaning* as well. " A is B " and " A is not B " are contraries so long as the meaning of the terms A and B remains the same, no matter what that meaning is; but the relation between " A is X " and " A is Y " can be determined only after we know the meaning of X and Y. If we learn that X and Y are opposites we can, of course, restate our second proposition, "A is Y," in the form " A is not X," the contrary of " A is X." If X and Y prove to be the same or similar in meaning the two original propositions are of course consistent.

Subject and predicate may both be different; *e. g.*, " Oxygen is heavier than nitrogen " and " Nitrogen is heavier than oxygen." Only one cannot be true; both might be false. If two propositions, alike in quality, have the same subject but have predicates which are complete opposites, as X and not-X, the propositions will be contradictories. Such pairs of terms as "rational and non-rational," "square and not-square," are examples.

Cases sometimes occur in which propositions having the same predicates but different subjects will be contradictory in meaning: thus, "Man alone is rational"; "Some being besides man is rational."

Similarly, propositions with unlike terms may stand in the relation of sub-contraries. For example: "Some men are rational" and "Some men are irrational," and "Simple substances make up a large part of the earth's crust" and "Compound substances make up a large part, etc.," are pairs of sub-contraries.

And likewise in the case of subalterns: for example, "All men are vertebrates" and "All men are mammals"; "No mental states can be weighed," "No emotions can be weighed." The relation of subaltern may hold between two propositions even if one of them is not universal. Thus, "Most books are worthless" and "Some books are worthless." Of course, "The recent novels are worthless," is not necessarily a subalternate of either of these propositions.

Singular propositions require special notice. "Socrates was the noblest of men" and "Socrates was not the noblest of men" are apparently contrary propositions, but as a matter of fact they are contradictories. Singular propositions which have the same terms are never contraries except in form. On the other hand, "Socrates was an Athenian" and "Socrates was a Spartan" are contraries.

Conversion.—A proposition is the verbal expression

of a judgment, and a judgment is an act of thought wherein we assert that certain relations hold among certain objects of thought, as, A is B, A is not B, some A is Y, and so on. Now it is often possible to formulate other propositions which are equivalent to these or which are obviously true if the original proposition be true. The proposition, "No conic sections are rectangular figures," is equivalent to "No rectangular figures are conic sections." The only difference between the two is in the *order of the terms.* The original subject and predicate have been interchanged. This process is known as Conversion. The proposition just converted was an E proposition and all E propositions can be converted.

Again, the proposition, "Some metals are elements" can be converted into "Some elements are metals." Both are I propositions. From the proposition, "Some quadrupeds are horses," we can get only "Some horses are quadrupeds." We happen to know that the same could be said of all horses, but we do not get that knowledge from the original proposition. The original statement is affirmative and affirmative propositions, as we have seen, do not give information about the whole of the predicate. But the statement, "All horses are quadrupeds," being universal, affirms something about the whole class horses.

The first general rule of conversion is that *no term may be distributed in the converse which was not distributed in the original proposition.* A proposition, then, such as "All A propositions are universal," can have as its converse only an I proposition, "Some universal propositions are A propositions."

Each of the propositions dealt with above has had as

its converse another proposition having the same quality. In all cases *the converse of a proposition must have the same quality as the original proposition.* This is a second rule of conversion. From this and the former rule it follows that the O proposition can have no converse. Its subject is undistributed and its quality negative; but in the converse, the original subject, having become the predicate of a negative proposition, would be distributed, a violation of the first rule.

	Proposition	Converse
The converse of A is I:	All S is P;	Some P is S.
The converse of I is I:	Some S is P;	Some P is S.
The converse of E is E:	No S is P;	No P is S.
O has no converse.		

The use of Euler's circles[5] will help to make these re-

A 

All S is P or (at least) some P is S.

E 

No S is P or no P is S.

I 

(At least) some S is P or some P is S.

O 

Some S is not P but all P may be S, or no P may be S, or some may be and some may not be.

lations clear. We see that propositions E and I are converted into E and I; in technical language they are converted "simply." A can be converted only into I;

[5] The forms here used are taken from Hyslop's *Elements of Logic.*

that is, it is converted into a proposition which is less general than itself, into a particular proposition. Such conversion is known as conversion by limitation or *per accidens.*

Obversion.—Again, we may find for all ordinary propositions an equivalent of the opposite quality: "All men are wise" is equivalent to "No men are unwise"; "Some men are just" can be expressed as "Some men are not unjust"; "No animals are moral" as "All animals are unmoral"; "Some men are not honest" as "Some men are dishonest." It will be noticed that in every case the subject is unchanged. (In "All men are wise," and "No men are wise," the subject is in both cases "All men"; the "No" belongs to the statement as a whole, not to the subject. The negative of "All men" would be "All who are not men.") This process is called *Obversion.* The *quality of the proposition* is changed and the *predicate of the obverse is the complete opposite of the predicate of the original proposition.*

	Proposition	Obverse
The obverse of A is E:	All S is P;	No S is not-P.
The obverse of I is O:	Some S is P;	Some S is not not-P.
The obverse of E is A:	No S is P;	All S is not-P.
The obverse of O is I:	Some S is not P;	Some S is not-P.

If the predicate in the original proposition be negative, as non-conductor, it will be replaced in the obverse by the corresponding positive term, conductor. "Some S is not-P" will have as its obverse "Some S is not P." Difficulty is likely to arise with regard to the predicate and its opposite. For example, the proposition, "No animals are moral," is not equivalent to "All animals are immoral." They may be neither moral nor immoral. The predicate in the new proposition must be *completely opposite* or contradictory to the original predicate. Sometimes it can not be expressed simply. For example, take the proposition, "The president is the nation's highest executive officer." In the obverse the *whole* of the predicate must be made negative, not simply "highest" or "executive" or "officer." It might read, "The president is not any one who is not the nation's highest executive officer."

The following symbols may be employed to indicate the relations considered in obversion. Suppose we have a proposition with the predicate P or not-P. Everything in the universe either has or has not the predicate P, that is, everything has one or the other of the predicates P and not-P. We may represent this fact by a circle divided into two compartments, thus:

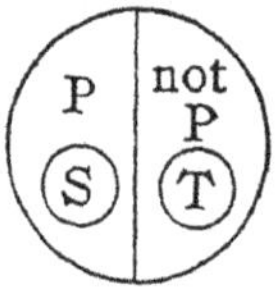

Then any given thing will fall into one or the other of those compartments. If our proposition asserts that it falls into one, that is tantamount to asserting that it falls outside the other; the latter assertion would be the obverse of the former. S is P, implies that S is not not-P; T is not-P, implies that T is not P.

Contraposition.—These changes in the form of propositions may both be present together and repeatedly. Let us take the proposition, "No men are immortal."

The obverse would be, "All men are mortal"; the converse of this, "Some mortals are men"; the obverse of this again, "Some mortals are not not-men" (not anything else than men); and this, being an O proposition, has no converse. We might, of course, have begun with conversion.

What is known as Contraposition is equivalent to obversion plus conversion.[6] The contrapositive of "No men are immortal" is "Some mortals are men"; the contrapositive of "All men are mortal" would be "No immortals are men." In the contrapositive the subject is the opposite of the original predicate, the predicate is the original subject, and the quality of the proposition is the opposite of that of the original proposition. An application of the rules for conversion will show that the I proposition has no contrapositive. The contrapositives of the various propositions are as follows:

	Proposition	Contrapositive
A, contrapositive E:	All S is P;	No not-P is S.
I, no contrapositive.		
E, contrapositve I:	No S is P;	Some not-P is S.
O, contrapositive I:	Some S is not P;	Some not-P is S.

In Conversion, Obversion, and Contraposition we have found certain variations in the forms in which a given thought content can be expressed. Such variations in the form of expression help to make clearer just what the content of the judgment really is. In these processes the changes in the form of expression are due to a change in the order of subject and predi-

[6] Some logicians add a second obversion. See Hibben, *Logic.*

cate or a change in the quality of predicate and copula, or both.

EXERCISES

1. Give contrary, contradictory, and subaltern of each of the propositions in Exercises "5," pages 62-65; where it is not possible to give all, state the reason why.

2. Classify the subjoined propositions into the four following groups:

1. Those which can be inferred from (1).
2. Those from which (1) can be inferred.
3. Those which do not contradict (1) but cannot be inferred from it.
4. Those which contradict (1).

(1) All just acts are expedient acts.
(2) No expedient acts are unjust.
(3) No just acts are inexpedient.
(4) All inexpedient acts are unjust.
(5) Some unjust acts are inexpedient.
(6) No expedient acts are just.
(7) Some inexpedient acts are unjust.
(8) All expedient acts are just.
(9) No inexpedient acts are unjust.
(10) All unjust acts are inexpedient.
(11) Some inexpedient acts are just acts.
(12) Some expedient acts are just.
(13) Some just acts are expedient.
(14) Some unjust acts are expedient. (Jevons.)

3. Give the converse, the obverse, and the contrapositive of each of the following propositions:

(1) All who were present were unprepared.
(2) No wise man would undertake such a task.
(3) Not to make the attempt is to confess yourself a coward.
(4) Charity begins at home.
(5) No statesman could have stooped to such a deed.
(6) Only a fanatic believes in panaceas.
(7) Discontent is frequently a symptom of inefficiency.

(8) A revolution is a surgical operation which self-appointed healers of social diseases are very ready to recommend as a preliminary to every cure.

(9) Uneasy rests the head which wears a crown.

(10) All organic substances contain carbon.

(11) Better late than never.

(12) Not many of the metals are lighter than water.

4. State the relation of each of the following propositions to the succeeding one:

(1) All the metals are elements.

(2) No metals are non-elements.

(3) No non-elements are metals.

(4) All non-elements are not-metals.

(5) All metals are elements.

(6) Some elements are metals.

(7) Some metals are elements.

(8) No metals are elements. (Hyslop.)

CHAPTER VIII *

THE SYLLOGISM

The Principles of Syllogistic Reasoning.—Let us return to the examination of the syllogism. Every complete syllogism contains three propositions and only three. They are a major premise, a minor premise and a conclusion. In the syllogism, "All men are mortal; Socrates is a man; therefore Socrates is mortal," the first proposition is the major premise, the second is the minor premise, and the third is, of course, the conclusion. The major premise is the broad foundation on which the reasoning rests. It is a universal proposition in syllogisms of this form. It may be either affirmative or negative. Thus we may have, "No men are immortal; Socrates is a man; therefore Socrates is not immortal." The major premise asserts that the whole of a certain class is included in another class or excluded from it, or it assigns a certain predicate to the whole of a certain subject.[1] The minor premise asserts that certain things are included in the first class; and the conclusion applies to these things the assertion which was made about the first class.

* This chapter may be omitted. The traditional treatment is given in chap. ix.

1 We have seen that all propositions may be regarded as stating a relation between classes, and that this way of regarding them is most useful and convenient. But other types of relations between subject and predicate exist and should not be forgotten.

Behind this reasoning lies the principle called the *Dictum de Omni et Nullo; i. e.,* "*Whatever statement may be made with regard to a class taken generally may be made of each and every member of that class,*" or "Whatever is true of each of the members of a class will be true of everything found to be a member of that class."

The application of this principle in the two syllogisms employed for illustration is perhaps obvious, but it may be well to represent the relation of the various terms graphically. For this purpose Euler's diagrams are valuable.

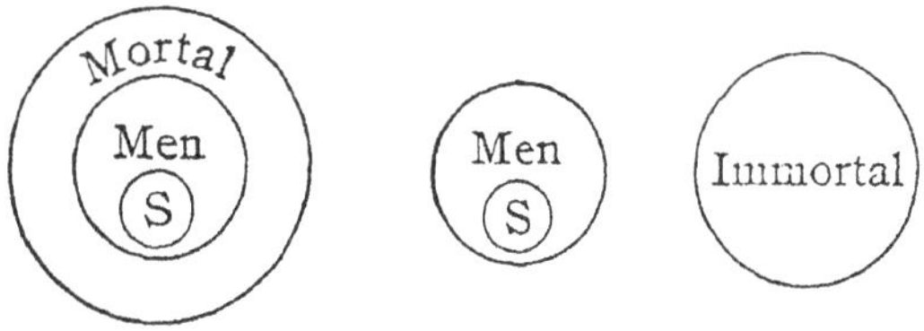

The class men is first included in the class mortal; the individual Socrates is then included in the class men and must in consequence be included in the class mortal. In the second case men is excluded from immortal and hence Socrates, who is included in men, will necessarily be excluded from immortal.

It will readily be seen that (1) *the minor premise in a syllogism of this sort cannot be negative.* A negative minor premise would assert that something did not belong to the class indicated by the subject of the major premise, and would give no ground for a further conclusion. The fact that something is true of a whole class of objects does not tell us whether it will or will not be true of some things not included in that class. Thus the premises, "All men are rational beings; monkeys are not men," do not warrant the conclusion that monkeys are not rational.[2] Other conclusions are

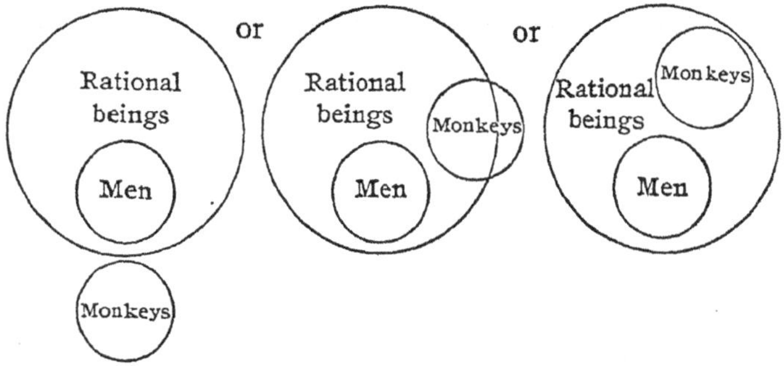

possible and we could not prove this one without using information not contained in our premises. If we tried to prove it by those premises alone, our reasoning would be invalid. An invalid syllogism is one in which the conclusion is not *proved* or *made necessary* by the premises. The conclusion may be true, but the function of the syllogism is to furnish a conclusion which *must* be true if the premises are true. Reasoning which does not prove the conclusion is fallacious, or in other words, it contains a fallacy.

We have seen that *the minor premise in a syllogism of this form cannot be negative.* It will be obvious that (2) *if the major premise be affirmative the conclusion must be affirmative* and that *if the premise be negative the conclusion must be negative.* If we affirm something of a whole class we cannot deny it of a part of the

2 These diagrams may be employed in illustrating the later rules also.

class, and if we deny it of the whole class we cannot assert it of a part. It is true also that (3) *the major premise cannot be particular.* If we have the premises, "Some animals can be domesticated; the wolf is an animal," we are obviously not justified in concluding that the wolf can be domesticated.

And again (4) *if the minor premise be particular, the conclusion cannot be universal;* it must be particular too. From the premises, "All works of art are valuable; some of the objects in this collection are works of art," we cannot conclude that all the objects in this collection are valuable. In no case can our conclusion contain more than was contained in the premises.

Syllogistic Proof.—In examining syllogistic reasoning the first question is not "Is the conclusion true?" but "*Does the conclusion follow necessarily from the premises?*" If the syllogism of this form correctly applies the *Dictum de Omni et Nullo* the reasoning is valid. A syllogism of this form is said to be in the *First Figure,* and this is the only form of syllogism which can be used to prove a universal affirmative proposition.

A Second Type of Syllogism.—There are several other varieties of syllogisms, each having certain special principles of its own. The one next to be discussed is used to prove that two facts or groups of facts are not the same; *it proves negative conclusions and only those.* Its Principle is this: *if one of two things is included in a class from which the other is excluded,*[3] *these things are excluded from each other.* To illustrate: "Every college-bred man has read that

[3] Or if one has a predicate which the other lacks.

book; this man has not read the book; therefore, this man is not college-bred." The fact that two subjects are included in the same class or that they possess the same predicate does not, on the other hand, prove that they are the same or that they are related in any other way. They might be even identical, it is true, but in the conclusion of a syllogism we are entitled to include *only what must be.*

Special rules.—In this sort of syllogism (1) *no conclusion can be drawn from two affirmative premises.* Either of the premises may be negative and one of them must be. In the illustration given above, the minor premise was negative. We might have "No college man would do this deed; this man has done it; therefore, he is not a college man."

On the other hand (2), *both premises can not be negative; only one may be negative.* If two things both lack the same predicate that fact alone does not warrant any further statement. "No Indians are Caucasians" and "No Chinamen are Caucasians" does not furnish any basis for a statement concerning the relation between Indians and Chinamen.

(3) *The major premise in a syllogism of this sort must be universal.* If our major premise were "Some college men are informed upon this subject," it would be possible that some were not, and the fact that a given man was ignorant of it would not prove that he was not a college man.

(4) *The minor premise may be either universal or particular; if it is particular, the conclusion must be particular; if it is universal, the conclusion may be universal.* It is obvious that if we make a statement about only a part of the class for which the subject of the

minor premise stands, we cannot make a statement about the whole of it in the conclusion. In every syllogism, the information contained in the conclusion must be furnished in the premises.

Major and Minor Premises.—It is not always easy to know immediately which is the major premise in a syllogism of this form. There is, however, a rule which can be applied to all syllogisms. *The major premise is the one which contains the major term. The major term is the predicate of the conclusion.* In an affirmative conclusion the subject may be regarded as contained in the predicate. In the proposition, "All triangles are geometrical figures," the class triangle is included in the class geometrical figures. The latter term is the major because it stands for the wider class. In negative propositions, the predicate is not necessarily wider; in the proposition, "No conic sections are triangles," the predicate is not wider than the subject. Still, for the sake of uniformity, the predicate of the conclusion is always called the major term. The syllogism first discussed is a syllogism in the *First Figure* and the other is in the *Second Figure.*

The Third Figure.—In the next type of syllogism, the subject is the same in both premises, but the predicates are different, or in other words, the subject is related by inclusion or exclusion to one class in the major premise and to another in the minor premise. The Principles of the Figure might be stated as follows:

1. *If a class (or individual) is included in each of two other classes those classes include each other, at least in part.* All X is Y and all X is Z; therefore, some Z is Y.

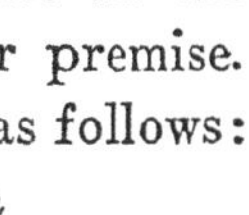

2. *If a class (or individual) is excluded from a*

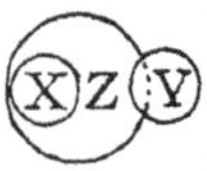

second class and included in a third, then a part, at least, of the third is excluded from the second. No X is Y and all X is Z; therefore, some Z is not Y.[4]

In some cases a conclusion is possible if only a part of the subject is described in one of the premises. Some of the special rules which follow state the conditions in which this is true. (1) *One of the premises must be universal.* For example, "No precious metals are soluble in sulphuric acid; some precious metals are soluble in nitric acid; therefore, some things soluble in nitric acid are not soluble in sulphuric acid." If the first makes an assertion about only a part of the class and the second likewise, we can not be certain that the two parts are the same and thus we learn nothing about the relation of their predicates. Thus, "Some triangles are scalene" and "Some triangles are right-angled" are premises which warrant no conclusion regarding the relation of scalene to right-angled figures. (2) *The major premise in this Figure may be either affirmative or negative. The conclusion will have the same quality as the major premise.* If "No A is B," and "Some (or all) A is C," then "Some C is not B"; or if "All A is B," and "Some A is C," then "Some C is B." (3) *The minor premise can not be negative.* If "All deer are herbivorous," and "No deer are hollow-horned animals," we cannot conclude that "No hollow-horned animals are herbivorous." (4) *The conclusion of a syllogism in the Third Figure is, in all cases, particular.* From "All men are mammals" and "All men are bi-

4 The student should test each of the special rules by the use of such diagrams.

peds" we can not conclude that "*All* bipeds are mammals." The fact that some or all of a certain class are included in another class or possess a given predicate (the minor premise asserts this) does give us information about a part of that predicate, but not about the whole (it does not distribute the predicate); as that predicate becomes the subject of the conclusion, the conclusion must be a particular proposition.

The Fourth Figure.—The three figures already discussed were described by Aristotle. The fourth is the invention of later logicians and is usually regarded as much less important than any of the others. In it the minor premise states something about the predicate of the major premise, and the conclusion in turn states something about the conclusion of the minor premise. Thus, "All great poems are the products of genius; all the products of genius are inimitable; therefore, some inimitable things are great poems." (If the conclusion were, "Great poems are inimitable," we should have a syllogism of the First Figure, and "All the products of genius, etc.," would be the major premise). The Principles of this Figure are: 1. *If a class is included in a second class and this in turn is included in a third, then the third will be partly coextensive with the first.*

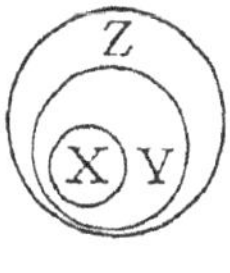

All X is Y
All Y is Z

Some Z is X

2. *If a class is excluded from a second and the latter is included in a third, then a part, at least, of the third will be excluded from the first.*

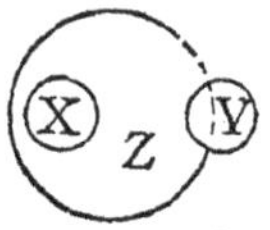

No Y is X
All X is Z

Some Z is not Y

3. *If a class is included in a second and the latter is excluded from a third, then the third will be excluded from the first.*

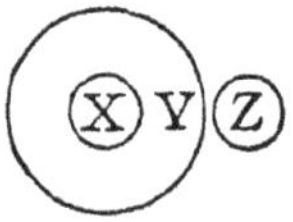

All X is Y
No Y is Z

No Z is X

In the illustration we have used it is obvious that we can not conclude that "All inimitable things are great poems." Our minor premise has not given us any information about the whole of the class, "inimitable things," so we can not have a universal conclusion in this instance. If our syllogism were, "All great poems are the products of genius; some products of genius are inimitable," it would be impossible to draw a conclusion. Poems might not happen to belong to the things included in the minor premise. The result would be similar if the minor premise were "Some products of genius are not inimitable." If the minor premise were, "No products of genius are inimitable," we could of course conclude that no inimitable things were poems. We can formulate this rule: (1) *If the major premise be affirmative the minor premise must be universal.* In all these instances, the major premise was the same and it was universal. Suppose we had "Some great poems were the products of genius." It will be seen that the minor premise, "All works of genius, etc.," will give a valid conclusion, but that none of the others will. (2) *If the major premise be affirmative and particular, the minor premise must be universal and affirmative.* With the minor premise "No products of genius are inimitable," no conclusion can be drawn; since only *some* great poems have been included in works of genius, it may well be that some inimitable things may be found among those not so included. (3) *The major premise may be negative.* "No great statesmen are selfish politicians; some (or all) selfish politicians amass great fortunes; therefore, some persons who amass great fortunes are not great statesmen." Some such persons might be great statesmen, so far as our premises are concerned; hence we have no right to conclude that no persons who amassed great fortunes were great statesmen.

EXERCISES

State the Figure and point out the errors in reasoning in the following syllogisms:

(1) All wisdom is desirable, but a knowledge of slang is not wisdom, and is, therefore, not desirable.

(2) Logic and mathematics furnish good mental training, and consequently the latter may be regarded as a branch of the former.

(3) Some athletes are susceptible to pneumonia, and as all these men are athletes some of them must be susceptible to pneumonia.

(4) Some industrious people are also bright, for there are both bright and industrious students in that group.

(5) Some statues are very lifelike, and no lifelike things are contrary to the laws of nature; hence, nothing contrary to the laws of nature is a statue.

(6) Some gymnastic exercises are good for increasing strength, but swimming is not, and hence is not a gymnastic exercise.

(7) All Democrats voted against the bill, and as most of our Congressmen are Democrats, they must all have voted against the bill.

(8) All M is P;
No M is S;
∴No S is P.

(9) Europeans cannot endure that climate; neither can Americans; hence, Americans may be regarded as a species of European.

(10) All ballads are interesting, and some interesting things are very old; hence, some very old things are ballads.

(11) All text-books are to be had at this store, but some novels are not to be had here, which proves that novels are not text-books.

For further examples see page 150 and page 177f.

CHAPTER IX

TRADITIONAL TREATMENT OF THE SYLLOGISM

The traditional treatment of the syllogism is simple though very formal. The syllogism is regarded as a form of reasoning in which each of two terms is compared with a third and as a result the two terms are found to be related to each other. Each of the two is compared with the third in a *premise*. The result of the comparison is stated in the *conclusion*.

	(M) is P	Premises.
	(S) is M	
∴	(S) is P	Conclusion.

P and S are found to stand in certain relations to M. In this case and in many others we are justified in asserting a relation between S and P: S and P are found to be related through M as a medium. For this reason M is called the *Middle Term* and the syllogism is said to embody *Mediate Reasoning*.

The validity of the reasoning is tested by the application of a number of rules. These rules have to do with the relation and distribution of the several terms in the syllogism. They are as follows:

1. Every syllogism contains three propositions and only three.

2. Every syllogism has three terms and only three. (If any term is ambiguous this rule is violated.)

3. The middle term must be distributed at least once.

4. No term may be distributed in the conclusion which was not distributed in one of the premises.

5. From two negative premises nothing can be inferred.

6. If one premise be negative, the conclusion must be negative; if both premises be affirmative, the conclusion must be affirmative.

7. From two particular premises no conclusion can be drawn.

8. If one premise be particular, the conclusion must be particular.

Let us examine these rules in the order given.

1. With more than three propositions, we should have more than a syllogism, though our reasoning might be valid.

2. The violation of rule two gives rise to the *Fallacy of Four Terms.* Unless two of the terms are confused this fallacy is not likely to arise. No one would try to draw a conclusion from the propositions, "Socrates was a philosopher," and "The earth revolves about the sun." But one might be tempted to draw a conclusion from the premises, "Steel is made from iron; iron is dug from the ground." Still, it would be wrong to conclude that steel is dug from the ground. The terms here are, "steel," "(something) made from iron," "iron," and "(something) dug from the ground."

3. The violation of rule three gives rise to the *Fallacy of Undistributed Middle.* Thus, the premises, "Some men are brave; and some men are strong," do not prove anything; nor do these: "All brave men

should be respected; and, all just men should be respected."

Let us represent the middle term by M, the minor term (subject of the conclusion) by S, and the major term (predicate of conclusion) by P. We are not justified by the premises in making any statement about the relation of S and P, for they may be wholly or partially identical or they may be mutually exclusive. But if the middle term were distributed we might be able to draw a conclusion. If all M is P and all S is M, we may conclude that all S is P.

Or if no M is P and all S is M, then no S is P.

An invalid syllogism is one in which it is not possible to determine fully the relation of the circles to each other, since there are conflicting possibilities. In the case of Undistributed Middle cited above, all, some, or none of S may be included in P.

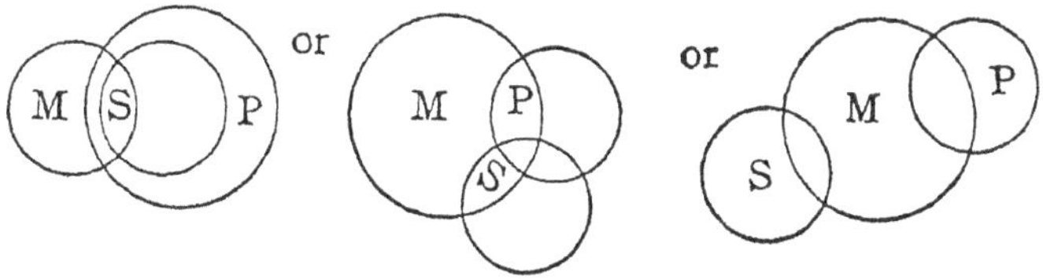

In valid syllogisms there may sometimes be a margin of indefiniteness (owing to the indefinite character of the "particular" propositions), but a certain amount of definite information regarding the relation of S and P is always given and the relation of the circles symbolizing major and minor terms is not left wholly in doubt.

The reason for the rule requiring the distribution of the middle term may be stated in this way: If each of two things is related to a part of a third, we can not conclude that they are related to each other, for they may not be related to the same part; but if one (or both) is related to the whole of the third, then it may be possible to assert a relation between the two. Thus:

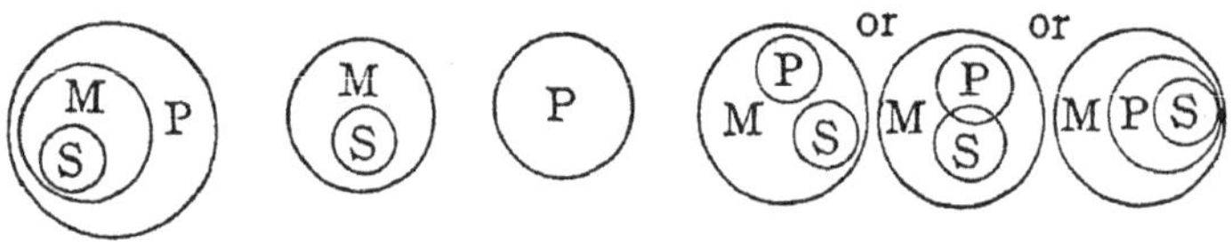

All M is P
All S is M
All S is P

No M is P
All S is M
No S is P

All P is M
All S is M
No conclusion.

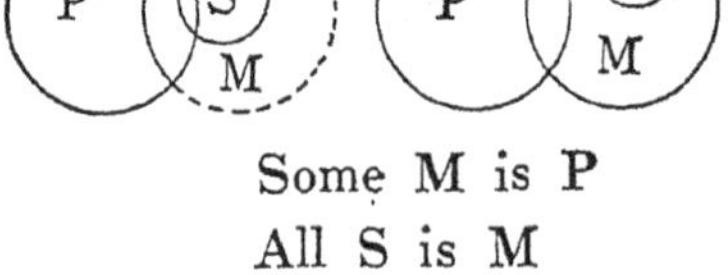

and so on

Some M is P
All S is M
No conclusion.

4. The reason for rule four is obvious: if we know something about only part of a class in the premise,

we can not say something about all of it in the conclusion. The violation of this rule gives rise to two fallacies: the *Illicit Process of the Major Term* and the *Illicit Process of the Minor Term.* In the syllogism, "All men (M) are vertebrates (P); all men (M) are rational (S); therefore, all rational beings (S) are vertebrates (P)," we have an illustration of the Illicit Process of the Minor Term, or Illicit Minor, as it is usually called. In the syllogism, "All Chinamen (M) are Mongolians (P); no Japanese (S) are Chinamen (M); therefore no Japanese (S) are Mongolians (P)," the Illicit Process of the Major Term occurs. Using the circles, we have for the first:

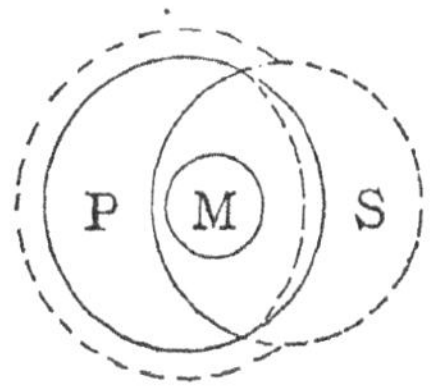

and for the second:

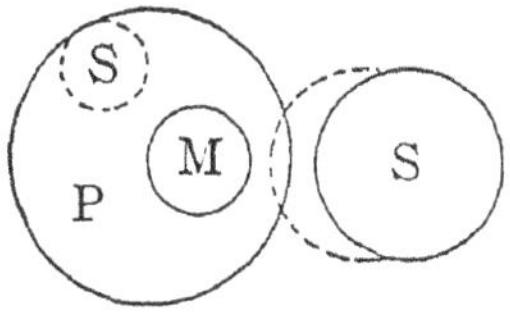

The dotted lines indicate possible boundaries of S.

The premises do not justify us in including all of S within the circle P in the first, nor of excluding all P from the circle S in the second. S may be outside of M, but still be wholly or partially within P.

5. With two negative premises, both major and

minor terms are excluded from the middle term, but that does not tell us whether they are or are not excluded from each other.

6. With one negative premise, either major or minor term is excluded from the middle term, while the other is not; therefore, if any relation can be asserted between major and minor terms, it must be one of exclusion.

7-8. The reasons for the two last rules can be more easily understood after we have considered the Moods and Figures of the syllogism. It will then be seen that the violation of these rules means a violation of rule three or rule four or both.

The Figure of a syllogism is determined by the positions of the middle term.[1]

The Four Figures are as follows:

1.	2.	3.	4.
M is P	P is M	M is P	P is M
S is M	S is M	M is S	M is S
∴S is P	∴S is P	∴S is P	∴S is P

In the First Figure, the Middle Term is the subject of the Major Premise and the predicate of the Minor Premise.

In the Second Figure it is the predicate of each.

In the Third Figure it is the subject of each.

In the Fourth Figure it is the predicate of the Major Premise and the subject of the Minor Premise.

The position of the Middle Term in the Third Figure

1 We have already seen that the Figures differ in other ways, but the traditional mode of distinguishing them is the one just mentioned.

is the opposite of that which it occupies in the Second; and in the Fourth it is the opposite of that in the First.

The Mood of a syllogism is determined by the quantity and quality of the several propositions which it contains. Propositions, as we have seen, are of four kinds with respect to quantity and quality, and are represented by the four letters, A, E, I, and O. The letters AAA would symbolize a syllogism in which each proposition was a universal affirmative. There are sixty-four possible moods:

AAA	~~AEA~~	~~AIA~~	~~AOA~~	~~EAA~~	~~EEA~~	~~EIA~~	~~EOA~~
~~AAE~~	AEE	~~AIE~~	~~AOE~~	EAE	~~EEE~~	~~EIE~~	~~EOE~~
AAI	~~AEI~~	AII	~~AOI~~	~~EAI~~	~~EEI~~	~~EII~~	~~EOI~~
~~AAO~~	AEO	~~AIO~~	AOO	EAO	~~EEO~~	EIO	~~EOO~~

~~IAA~~	~~IEA~~	~~IIA~~	~~IOA~~	~~OAA~~	~~OEA~~	~~OIA~~	~~OOA~~
~~IAE~~	~~IEE~~	~~IIE~~	~~IOE~~	~~OAE~~	~~OEE~~	~~OIE~~	~~OOE~~
IAI	~~IEI~~	~~III~~	~~IOI~~	~~OAI~~	~~OEI~~	~~OII~~	~~OOI~~
~~IAO~~	(IEO)	~~IIO~~	~~IOO~~	OAO	~~OEO~~	~~OIO~~	OOO

Many of these are at once seen to be invalid: thus, applying the rules for negative and particular premises, we can eliminate those moods through which a line is drawn. The mood IEO does not violate any of those rules, but examination will show that it will give a fallacy of Illicit Major in each of the Figures. The conclusion is negative and hence distributes its predicate, the major term. The major premise is an I proposition and hence distributes neither of its terms; therefore, the major term can not be distributed in the premise, and hence this mood may be eliminated also.

There remain only eleven moods which may be valid, but many of those are invalid in some of the figures. We will examine each of these moods in each of the figures.[2]

In the First Figure we should have the following results:

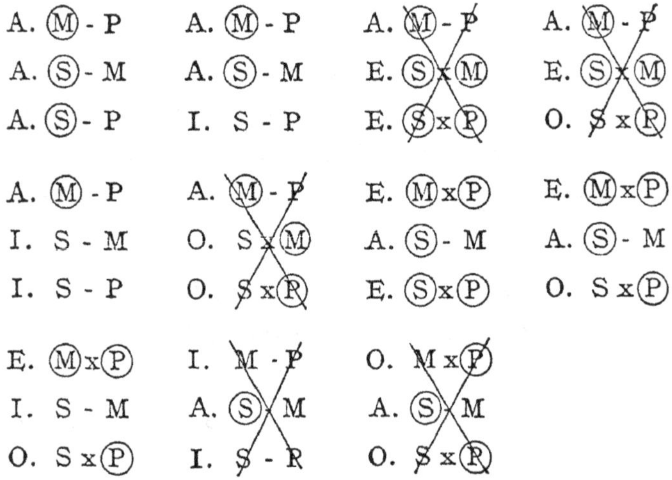

It is evident that the following are invalid in the First Figure: AEE, AEO, AOO, IAI, and OAO. IAI and OAO are invalid because of an Undistributed Middle, the others because of Illicit Majors.

The valid moods are AAA, AAI, AII, EAE, EAO, and EIO. AAI and EAO are necessarily valid since AAA and EAE are valid. I and O are called weakened conclusions because they are less general than they

[2] To facilitate dealing with them we shall employ the symbols used in the exposition of Conversion and Obversion.

might be. A comparison of these moods will show that two general statements may be made regarding reasoning in this figure:

1. *The Major premise must be universal.*
2. *The minor premise must be affirmative.*

Examination of the illustrations given in the previous discussion of this Figure (page 127) will show that every syllogism which violated either of these two rules failed to give a valid conclusion.

In the Second Figure the results are different:

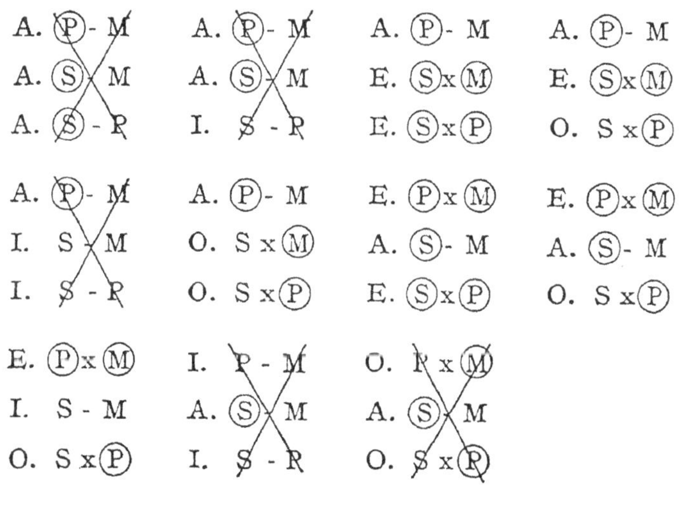

Here the moods, AAA, AAI, AII, IAI, and OAO are invalid, the last because of Illicit Major, the others because of Undistributed Middle.

The valid moods are AEE, AEO, AOO, EAE, EAO, and EIO. O is a weakened conclusion in AEO and EAO. Here we find that in the Second Figure:

1. *The major premise must be universal.* 2. *One premise must be negative and the conclusion likewise must be negative.*

These rules, like those of the First Figure, might have been formulated on the basis of the typical cases presented in the earlier discussion. (See p. 130.)

Again in the Third Figure:

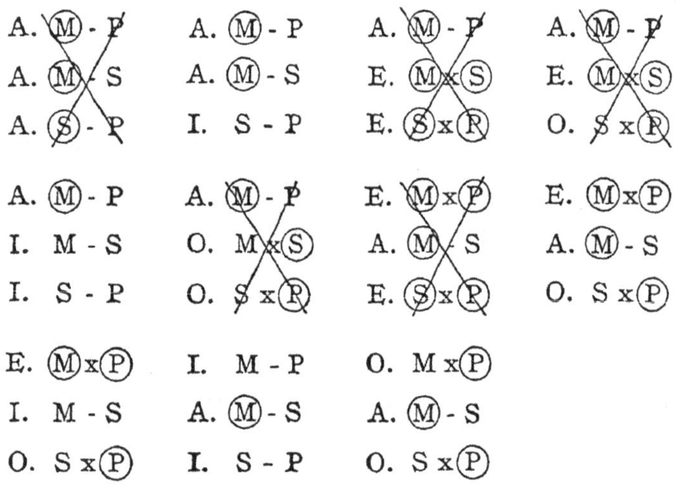

In this case the invalid moods are AAA, AEE, AEO, AOO, EAE. Of these, AAA and EAE are cases of Illicit Minor; and the rest, of Illicit Majors. The valid moods here are AAI, AII, EAO, EIO, IAI, and OAO. In the Third Figure:

1. *The conclusion must be particular.*

2. *The minor premise must be affirmative.*

In this case, as in the others, the rules might have been discovered, without consideration of the moods, by a direct examination of cases.

In the Fourth Figure we have:

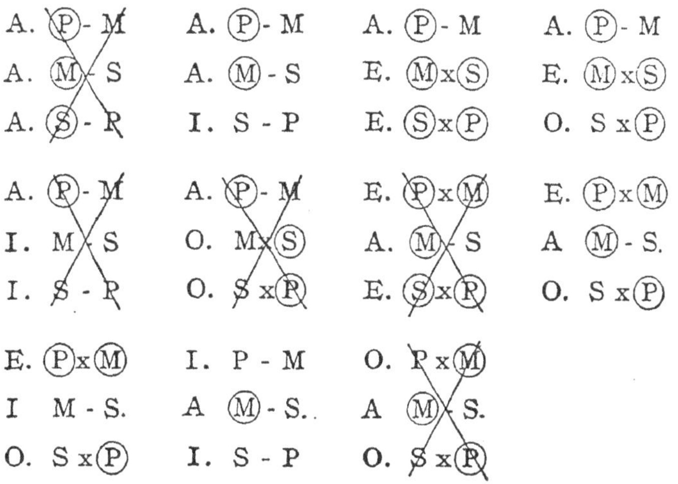

Here the invalid moods are AAA, AII, AOO, EAE, and OAO. AAA and EAE give Illicit Minor, AII and AOO give Undistributed Middle, and OAO gives Illicit Major. The valid moods are AAI, AEE, AEO, EIO, IAI. We get these rules for the Fourth Figure:

1. *If the major premise be affirmative, the minor premise must be universal.*

2. *If the major premise be also particular, the minor premise must be affirmative.*

3. *If the minor premise be affirmative, the conclusion must be particular.*

4. *If either premise be negative, the major must be universal.*

5. *The conclusion may not be a universal affirmative proposition.*

Comparison of all the valid moods shows that the mood AAA is valid in the First Figure only. As this is the only mood in which A appears as a valid conclusion, it will be evident that a universal affirmative conclusion can be proved in the First Figure only.

REDUCTION OF THE MOODS AND FIGURES.

The Medieval schoolmen invented a set of menemonic verses to serve as an aid to the memory in recalling the valid modes in the several Figures. The verses consisted of barbarous Latin terms. The words contain also letters for guidance in reduction of the other Figures into the First. These verses, with their interpretation, are as follows:

> B*a*rb*a*r*a*, C*e*l*a*r*e*nt, D*a*r*ii*, F*e*r*io*que prioris;
> C*e*s*a*r*e*, C*a*m*e*str*e*s F*e*st*i*n*o*, B*a*r*o*k*o* secundae;
> Tertia, D*a*r*a*pt*i*, D*i*s*a*m*i*s, D*a*t*i*s*i*, F*e*l*a*pt*o*n,
> B*o*k*a*rd*o*, F*e*r*i*s*o*n, habet, quarta insuper addit
> Br*a*m*a*nt*i*p, C*a*m*e*n*e*s; D*i*m*a*r*i*s, F*e*s*a*p*o*, Fr*e*s*i*s*o*n.

The moods are indicated by the italicised letters. All the valid moods are included except those in which there are so-called weakened conclusions, *i. e.*, cases in which a particular conclusion is drawn, though a universal would be valid, such as AAI or EAO in the First Figure. The first line indicates the moods of the First Figure, the second line, of the Second, the third and the first half of the fourth indicate those of the Third Figure, and the last line, those of the Fourth Figure.

The First Figure was regarded as the Perfect Figure, and the others were transformed into it by making certain changes in their various members. This process was called the Reduction of the Imperfect Figures. These words contain letters which stand for the changes which must be made. The capital letters in the last four lines indicate the mood of the First Figure to which the mood, indicated by the word in which they are found, may be reduced. Thus Cesare may be reduced to Celarent. p indicates that the preceding proposition is to be converted *per accidens* or by limitation; s indicates that the preceding proposition is to be converted simply, and m indicates that the premises are to be transposed.

CAMESTRES	CELARENT
All A is C	C is not B
(All stars are suns)	(No suns are planets)
No B is C	All A is C
(No planets are suns)	(All stars are suns)
Therefore, B is not A	Therefore, no A is B.
(No planets are stars)	(No stars are planets)

The minor premise in Camestres is first converted, then the two premises are transposed, and finally the conclusion is converted. To reduce Cesare to Celarent we need only convert the major premise.

As a second example we may take the reduction of Bramantip to Barbara.

Bramantip	Barbara
All C is B	All B is A
All B is A	All C is B
Some A is C	All C is A

In this case, the premises are transposed, and the conclusion is converted. This would give AAI. But the conclusion A would be valid from these premises. The p in this case may be taken as indicating that, instead of a conclusion less in quantity than the original proposition, we may have one which is greater in quantity, namely, universal.

There is one more significant letter in these words, the letter k. It indicates that the reduction must be made by indirect means. Take, for example, Bokardo which reduces to Barbara.

Bokardo	Barbara
Some A is not C	All B is C
All A is B	All A is B
Therefore, some B is not C	Therefore, all A is C

In this case the major premise of Bokardo is the contradictory of the conclusion of Barbara; and the conclusion of Bokardo is the contradictory of the major premise of Barbara. Suppose the conclusion of Bokardo to be false; then its contradictory, "All B is C," will be true; taking this as the major premise of a new syllogism and the proposition, "all A is B," as the minor premise, the conclusion will be the contradictory of the major premise of Bokardo and the new syllogism will be in the mood Barbara. Bokardo may be also be reduced to Darii. First obvert, then convert, the major premise; transposing the two premises, we then have Darii. All this mechanism is entirely unscientific and its interest is purely historical.

EXERCISES ON THE SYLLOGISM

1. What kinds of propositions are incapable of proof in the Second, Third and Fourth Figures respectively? Give the reasons for your reply.
2. If either premise of a syllogism is O, what must the other be?
3. With I as the major premise, what must the minor premise be?

4. Show that an E proposition is highly efficient as a major premise. (J.)[3]

5. Show that O is seldom admissible as a minor premise. (J.)

6. Prove that there must always be in the premises one more distributed term than in the conclusion. (J.)

7. Prove from the general rules of the syllogism, that when the major term is the predicate in its premise, the minor premise must be affirmative. (J.)

8. Point out which of the following pairs of premises will give a syllogistic conclusion, and name the obstacle which exists in other cases.

(1) No A is B; some B is not C.
(2) No A is B; some not C is B.
(3) All B is not A; some not A is B.
(4) Some not A is B; no C is B.
(5) All not B is C; some not A is B.
(6) All A is B; all not C is B.
(7) All not B is not C; all not A is not B.
(8) All A is not B; no B is C.
(9) All C is not B; no A is not B.

3 (J) refers to Jevons, *Studies in Deductive Logic,* where a great many more questions of this character may be found. The following exercise is from the same source.

CHAPTER X

ABBREVIATED AND COMPLEX FORMS OF REASONING — HYPOTHETICAL AND DISJUNCTIVE SYLLOGISMS

The Enthymeme.—Usually our reasoning does not fall into the form of a perfect syllogism. In the first place it very often happens that one or another of the propositions is omitted. For example, "This object can be magnetized, for it is made of iron," omits the major premise, "All things made of iron can be magnetized." Again, in "Every member of the jury voted for acquittal, therefore X voted for acquittal," the minor premise, "X was a member of the jury," is omitted. In "All metals are elements; this is a metal," the conclusion is omitted.

Syllogisms from which one proposition is missing are called Enthymemes. The missing premise can usually be found without difficulty. The two propositions which are given contain the three terms of the syllogism; one of these will be common to the two propositions, and the missing proposition will contain the other two terms. Thus with the proposition, "S is M; hence, S is P," the missing premise is clearly "M is P," or "P is M." With "M is P; therefore, S is P," the missing premise will contain S and M.

The danger of false reasoning is greater here than in the complete syllogism, since the proposition which is not expressed may be false or inadequate, and if the proposition is not definitely stated its inadequacy is easily overlooked.

The Enthymeme is an incomplete form of syllogistic reasoning; it is less than a syllogism. There are several complex forms in which we find more than a syllogism.

PROSYLLOGISM AND EPISYLLOGISM.—Two complete syllogisms may be united by having a proposition in common. Thus:

Prosyllogism { All the Romance languages are derived from Latin;
French is a Romance language;
Therefore, French is derived from Latin. }

Episyllogism { This man speaks French;
Therefore, this man speaks a language derived from Latin. }

In this example the conclusion of the first syllogism is the major premise of the second. This is known as Prosyllogism and Episyllogism, the conclusion of the Prosyllogism being the major premise of the Episyllogism. One syllogism might, of course, establish the minor premise of the other:

Prosyllogism { French is a Romance language;
This man speaks French.
Therefore, this man speaks a Romance language. }

Episyllogism { All the Romance languages are derived from Latin;
Hence, this man speaks a language derived from Latin. }

Again, it might have each of its premises established by another syllogism:

Prosyllogism:
- Everything which is able to restrain trade is a source of danger;
- Every monopoly is able to restrain trade;
- Hence, every monopoly is a source of danger.

Prosyllogism:
- A company which has complete control of a certain commodity is a monopoly;
- This trust has complete control of a certain commodity;
- Hence, this trust is a monopoly.

Conclusion: Therefore, this trust is a source of danger.

An enthymeme might take the place of the complete syllogism in the case of either or both of the prosyllogisms. Further, the premises of the prosyllogisms might themselves be supported by other syllogisms.

A great many syllogisms may be combined into one reasoning process, and most reasoning processes contain several syllogisms, complete or abbreviated.[1]

[1] Geometrical reasoning illustrates abbreviated reasoning very clearly. For example take the proof of the proposition that "All straight angles are equal."

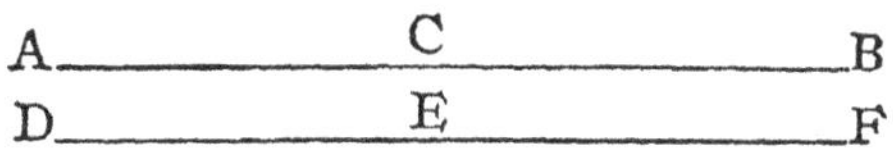

"Let the angles ACB and DEF be any two straight angles. To prove that the angle ACB equals the angle DEF.

"Place the angle ACB on the angle DEF, so that the vertex C shall fall on the vertex E, and the side CB on the side EF. Then CA will fall on ED. Therefore the angle ACB equals the angle DEF." (Wentworth, *Plane Geometry*, page 14.)

This is the proof in an abbreviated form. It might be more fully expressed as follows: (*See page* 154.)

We might have a chain of syllogisms in which the conclusion of each was the minor premise of the one following.

All ungulates are mammals.
All mammals are warm-blooded.
All ungulates are warm-blooded.

All warm-blooded animals have lungs.
All ungulates are warm-blooded animals.
All ungulates have lungs.

All animals that have lungs require air.
All ungulates have lungs.
All ungulates require air.

What is true of the angles ACB and DEF will be true of all straight angles.

Two angles which can be so placed upon each other that their vertices coincide and their sides coincide are equal, each with the other.

Any figure may be moved from one place to another without altering its shape. (Axiom of superposition.) Therefore, the figure ACB may be placed upon the figure DEF without altering its shape.

Straight angles are such as have their sides extending in opposite directions so as to be in the same straight line. The angles ACB and DEF have their sides so extending. Hence the lines AB and EF are straight lines.

Two straight lines which have two points in common coincide and and form but one line.

When the figure ACB is superposed on the figure DEF so that the vertex C shall fall on the vertex E, and the side CB on the side EF, the straight line AB falls on the straight line DF and they coincide; the line CA falls on the line ED, and coincides with it, CB coincides with EF, [And C coincides with E].

Therefore the angle ACB and the angle DEF are equal.

Therefore all straight angles are equal.

Geometrical reasoning is not all *syllogistic* in the narrowest sense of the word. See chapter xvii.

THE SORITES.—Now, instead of putting the conclusion in words and repeating it in the succeeding proposition, we may omit everything except the new premises until we are ready to draw the final conclusion. Thus:

	All ungulates are mammals.	A is B
	All mammals are warm-blooded.	B is C
	All warm-blooded animals have lungs.	C is D
	All animals that have lungs require air.	D is E
Hence,	All ungulates require air.	A is E.

This is known as the *Sorites;* a Sorites may have any number of members. There are two forms. That given above, is an example of the Progressive or Aristotelian Sorites. The premise containing the subject of the conclusion (the Final Minor) comes first in order; that containing its predicate (The Prime Major) comes last; the intermediate propositions serve to connect the two. In the Regressive or Goclenian Sorites, the Prime Major comes first and the Final Minor last among the premises. If expanded into a chain of prosyllogisms and episyllogisms, the conclusion of each syllogism would be the major premise of the one following. For example:

	A European is a Caucasian.	B is A
	A Frenchman is a European.	C is B
	A Parisian is a Frenchman.	D is C
	This author is a Parisian.	E is D
Hence,	This author is a Caucasian.	E is A

In both forms of the sorites the reasoning is in the first figure of the syllogism. With the exception of the terms which are contained in the conclusion, every term in the sorites is a middle term. The greatest source of danger in this form of reasoning is to be found in ambiguous terms.

Only the Final Minor premise may be particular; only the Prime Major may be negative.

Hypothetical Reasoning.—The forms of reasoning with which we have been dealing in the last three chapters have employed only declarative sentences, or *Categorical Propositions*, as they are called in Logic. A categorical proposition is an unconditional statement. "A is B" or "A is not B" are typical forms. But there are other kinds of propositions; one of these is the *Hypothetical Proposition.* A hypothetical proposition is one containing a categorical proposition and the statement of a condition on which the truth of the categorical depends. The conditional member of the proposition is called the *Antecedent;* the categorical member is called the *Consequent.* A hypothetical proposition may be made the major premise of a syllogism.

Such a syllogism would be a HYPOTHETICAL SYLLOGISM. The Hypothetical syllogism has four forms.

1. If A is B, C is D.	If the substance is carbon, it will burn.
A is B	It is carbon.
∴ C is D	∴ It will burn.
2. If A is B, C is D.	If the substance is carbon, it will burn.
A is not B	It is not carbon.
∴ C is not D	∴ It will not burn.
3. If A is B, C is D.	If the substance is carbon, it will burn.
C is D	It will burn.
∴ A is B	∴ It is carbon.
4. If A is B, C is D.	If the substance is carbon, it will burn.
C is not D	It will not burn.
∴ A is not B	∴ It is not carbon.

The first of these affirms the antecedent, the second denies it; the third affirms the consequent, and the fourth denies it. The second and third are obviously invalid. The fact that the substance is not carbon gives us no further information about qualities; and the fact that it will burn does not insure its being carbon. These instances are typical and illustrate the general rule that *Denying the antecedent or affirming the consequent in a hypothetical syllogism are invalid forms of reasoning.*[2]

We may have a hypothetical syllogism in which the minor premise is also a hypothetical proposition.

If A is B, C is D.	If he is nominated, he will be elected.
If C is D, E is F.	If he is elected, this measure will not pass.
∴ If A is B, E is F.	∴ If he is nominated, this measure will not pass.

Disjunctive Reasoning.—There is a third kind of so-called syllogism with still another sort of proposition as its major premise. This is the *Disjunctive Syllogism* and *its major premise is a Disjunctive Proposition.* A disjunctive proposition is one which states an alternative. "A is either B or C"; "It will either rain or snow." The minor premise either affirms or

2 There are cases, however, in which these forms give true conclusions. If the antecedent is the only one on which the consequent would follow then all the forms of the hypothetical syllogism would give valid conclusions. For example, if we had the major premise, "If A is B, and in no other case, C will be D," then to deny that A is B would necessitate the conclusion that C is not D. We may take, as a concrete case, "If a triangle is equilateral, and in no other circumstances, it will be equiangular. This triangle is not equiangular; hence it is not equilateral"; or, "This triangle is equilateral, therefore it is equiangular," and so on.

denies one of the alternatives. The conclusion either denies or affirms the other.

A is either B or C	It will either rain or snow.
A is B	It will rain.
∴ A is not C	∴ It will not snow.
A is either B or C	It will either rain or snow.
A is C	It will snow.
∴ A is not B	∴ It will not rain.
A is either B or C	It will either rain or snow.
A is not B	It will not rain.
∴ A is C	∴ It will snow.
A is either B or C	It will either rain or snow.
A is not C	It will not snow.
∴ A is B	∴ It will rain.

All these forms are valid. The only source of danger is in the major premise. If the alternatives are not true alternatives, the conclusion can not be trusted. If A can be anything else than B or C, or if it can be both at the same time, the denial or affirmation of one alternative cannot assure us of the truth or falsity of the others.

There are more complex forms of disjunctive reasoning; we might, for example, have the proposition, " A is B or C or D, etc." In this case the affirmation of one would mean the denial of the other two; but the denial of one would give as the conclusion a disjunctive proposition containing the two others as alternatives. Thus " A is not B; A is either C or D, etc." Similarly the assertion " A is B or C " would give the conclusion " A is not D, etc.," and so on.

There are certain imperfect forms of this syllogism which are sometimes useful. Sometimes we know that A is B or C or, it may be, both. In such a case, if we know that A is not B, then it must be C, but if we know that it is B, we do not know that it is not also C. With such a major premise, the minor premises which are affirmative do not give valid conclusions. It would be simpler in such cases to state the three possibilities as mutually exclusive, " A is B or C, or both B and C," and proceed as in the perfect forms of the hypothetical syllogism.

More Complex Forms. THE DILEMMA.—There are more complex forms of reasoning in which hypothetical and disjunctive propositions are combined. Thus we may have:

If A is B, C is D or E (*or* C is D or E is F).
A is B. ∴ C is D or E (*or* C is D or E is F).

More concretely:

If he fails, he will leave college or drop back a class.
But he is sure to fail.
∴ He will leave college or drop back a class.

If the antecedent were denied there could be no valid conclusion; in this and all other respects this syllogism is like a simple hypothetical syllogism except in having a disjunctive consequent, instead of a categorical one. We get more complicated forms when the major premise consists of two hypothetical propositions, in which either the antecedents or the consequents are found to be alternative: the minor premise is a disjunctive proposition, and the resulting syllogism is a Dilemma.

> If A is B, C is D; and if E is F, C is D.
> But either A is B or E is F.
> ∴ C is D.

If a college education gives a student useful information, it is valuable to him.
If it gives him mental training it is valuable to him.
But it either gives him useful information or mental training.
∴ It is valuable to him.

This is a *Simple Constructive Dilemma:* simple because the consequents of the hypothetical propositions in the major premise are the same in both cases; constructive because it establishes an affirmative conclusion.

If the consequents were denied we should not have a dilemma, but two simple hypothetical syllogisms. There would be no disjunctive premise.

The second form of the dilemma is the *Complex Constructive Dilemma.* In this, the consequents of the hypothetical propositions in the major premise are not the same.

> If A is B, C is D; and if E is F, G is H.
> But either A is B or E is F.
> ∴ Either C is D, or G is H.

"If a statesman who sees his former opinions to be wrong does not alter his course he is guilty of deceit; and if he does alter his course he is open to a charge of inconsistency; but either he does not alter his course or he does; therefore, he is either guilty of deceit, or he is open to a charge of inconsistency." (Jevons, *Lessons in Logic,* p. 168.)

Unlike the simple dilemma, this has a disjunctive conclusion.

The *Complex Destructive Dilemma* has a negative minor premise and a negative conclusion.

If A is B, C is D; and if E is F, G is H.
But either C is not D or G is not H.
∴ Either A is not B or E is not F.

"If this man were wise, he would not speak irreverently of the Scripture in jest; and if he were good he would not do so in earnest; but he does it either in jest or in earnest; therefore, he is either not wise, or not good. (Whately, *Elements of Logic.*)

If the minor premise were "Neither C is D nor G is H" we should not have a dilemma. The minor premise would not be disjunctive and we should have two simple hypothetical syllogisms.

Asserting that one or the other of the antecedents was false, or that one or the other of the consequents was true would be fallacious, as in the case of the simple hypothetical syllogism.

In practice it is very difficult to find true major premises for a dilemma. Moreover, "a dilemma can often be retorted by producing as cogent a dilemma to a contrary effect. Thus an Athenian mother, according to Aristotle, addressed her son in the following words: "Do not enter into public business; for if you say what is just, men will hate you; and if you say what is unjust, the gods will hate you." To which Aristotle suggests the following retort: "I ought to enter into public affairs; for if I say what is just, the gods will love me; and if I say what is unjust, men will love me." (*Jevons.*)

The conclusion of a dilemma, as of any other form of reasoning, may serve as a premise for further reasoning.

Extra-syllogistic Reasoning.—Certain other forms of reasoning call for some discussion here. They are not syllogistic, but they are closely related to syllogistic reasoning. For example, "A is taller than B; B is taller than C; therefore A is taller than C." This is not a syllogism. There are five terms in the reasoning; A, B, C, taller than B, and taller than C. There is a similar difficulty in this: "M is east of N; N is east of O, therefore M is east of O." It would, of course, be possible to construct a syllogism which would cover the ground in each of these cases. Thus, "Whatever is taller than another thing is taller than everything which is shorter than that thing; A, B, and C present a case, etc." And similarly in the other example. Some such principles as these are implied, but the reasoning as stated is not in the form of a simple syllogism. We have here a system of relations which is more complicated than that found in the ordinary syllogism. In the latter we need have only our premises and the ordinary laws of reasoning, to assure our conclusion; in reasoning of the sort illustrated in these examples we must have, besides our premises, a supply of information about the general system of things to which the data in question belong. When such reasoning is thrown into the syllogistic form the major premise states the main principles of the system, as in the example above. Sometimes our information about the system would be sufficient to warrant a conclusion and sometimes it would not; sometimes the fact that two things are related to a third gives us information regarding their relations to each other and sometimes it does not. From the statements that "A is the employer or friend or enemy of B" and that "B is the employer or friend or enemy of C," we could not draw any conclusion with regard to A's direct relations to C. Things equal to the same thing are equal to each other, but it is not necessarily true that things unequal to the same thing will be unequal to each other. In the latter case there are two possibilities; the system is not clearly enough defined to make certain any conclusion at all. This was true in some of the illustrations given above. To decide in any given case we must first determine whether the set of relations involved is completely enough known to justify a conclusion; in other words, Is more than one conclusion possible? Sometimes the system may be very complex, but its parts may be so related to each other and so completely known as to make a conclusion possible. A conclusion may simply state the relation of the terms in a reverse order. "A is east of B, therefore B is west of A." The conclusion may represent a pathway through a system from one particular part to another; while the premise may be merely the same pathway followed from the other end. If A is the son-in-law of the half-sister of B's grandfather, then B is the grandson of the half-brother of A's mother-in-law. The system might have any degree of complication whatever, but if the several relations could be read from either end the pathway could be followed in either direction. Reasoning of this sort might be regarded as a broader form of conversion.[3]

[3] See Aikins, *Principles of Logic,* chap. xi.

EXERCISES.

1. Supply the missing propositions in the following:
   (1) He is a politician and therefore not to be trusted.
   (2) They were all brave men and this man was one of them.
   (3) Whales have warm blood but fish do not.
   (4) Only members will be admitted; that excludes you.

2. Determine which of the following give valid conclusions and which do not; point out the fallacies involved:
   (1) If he goes, I shall remain; but he will not go.
   (2) If he goes, I shall remain; and I shall remain. main.
   (3) I shall remain if he goes; and he will go.
   (4) If it rains to-morrow, the game will be postponed; the game will be postponed.
   (5) If all the sides of this triangle are equal its angles are equal too; now its angles are not equal.
   (6) If he fails it will be because he has not worked hard; and he has not worked hard.

3. Criticise the following, stating the form of reasoning in each case:
   (1) A great man must either have extraordinary natural ability or exceptional capacity for work; this man had extraordinary natural ability, hence we may assume that his capacity or work was not unusual.
   (2) If the government enacts such a law it must either adopt socialism or go into bankruptcy; but it will not enact such a law; so there is no danger of either socialism or bankruptcy.—(Hyslop.)
   (3) If capital punishment involves cruelty to its victims it ought to be abolished in favor of some other penalty; if it does no good for society it should also be abolished. But either it involves cruelty to its victims or it does no

good to society, and hence it ought to be abolished.—(Hyslop.)

(4) If he sinks he will be drowned, and if he swims he will be captured by the enemy; but he must either sink or swim; therefore, he will either be drowned or captured by the enemy.

(5) If he tells the truth he will be forgiven, and if he does not he will escape detection; but either he will be forgiven or he will escape detection; hence he will either tell the truth or he will not.

(6) If he did that intentionally he is not wise, and if he did it unintentionally he is not lucky; but he is neither wise nor lucky; therefore, he did it neither intentionally nor unintentionally.

4. In a sorites why must all the premises except the prime major be affirmative and all except the final minor be universal?

CHAPTER XI

I. PROOF AND DISPROOF. II. FAILURE TO PROVE

I. **Various Kinds of Proof.**—The conclusion of a valid syllogism is *proved;* and so also is the conclusion of each of the other forms of reasoning which we have examined. A proposition is proved when it is shown to be the necessary consequence of any combination of admitted propositions. All the cases which we have so far examined are instances of *direct proof.* In direct proof we show that, granted certain things, the conclusion necessarily follows. A conclusion is proved only when it is shown that the conclusion *must* be true.

There are several other kinds of proof; in this chapter we shall consider the other kinds, and also the various forms of failure to prove, or fallacy.

INDIRECT PROOF.—The first to be considered is *indirect proof;* we prove a proposition indirectly by *disproving* its contradictory, *i. e.*, by showing that its contradictory cannot be true.

To disprove a proposition it is necessary to find some admitted fact or truth which is inconsistent with it. For instance, if we can show that the contrary or contradictory of a proposition is true, the proposition must be false. More concretely, if we have an A proposition, the contradictory would be an O proposition; if we can show that O is true, A is necessarily false, and to show the truth of O we need find only one real exception to A. Showing the truth of E would also disprove A; but E is a universal proposition, and it is obviously much more difficult, in ordinary circumstances, to prove a universal than it is to prove a particular. Similarly the truth of I or of A would mean the falsity of E, etc.

In indirect proof, we *disprove the contradictory*, not the contrary, for the falsity of the *contrary* does not prove the truth of the proposition, since both contraries may be false; but if the *contradictory* is false, the proposition must be true, for one of two contradictories must be true. Suppose then that we wish to prove an A proposition: if we can show, in any way, that the corresponding O proposition would be false or absurd —contrary to fact or reason—our thesis is proved.

The contradictory is usually disproved by showing that some of its necessary consequences are absurd.

Indirect proof is frequently employed in geometry and it is there that the best examples of it are to be found. It is also a frequent resource in political debate, but in that field the facts are so complicated and the matter of establishing any proposition so liable to error that the grounds of any conclusion established in this way must be very carefully examined.

We shall consider briefly two other special forms of proof, which are perhaps reducible to direct proof, but they are apparently very different from what we find in the syllogism and they will therefore be considered separately. The first is found in geometrical reasoning.

PROOF IN GEOMETRY.—In geometrical proof we seem

to be founding a universal conclusion upon a single case; how is it possible for us to have perfect confidence in a conclusion which seems at first sight to be an induction from one isolated figure? If the individual peculiarities of the figure had anything to do with supporting the conclusion the latter would of course be of very slight value; we should have no assurance that the next example might not be inconsistent with the conclusion. The certainty of the conclusion rests upon the fact that the figure used in the demonstration is, in all essential respects, like every other figure to which the proof is supposed to apply. The figure employed is purely symbolical; it stands for certain universal relations. If the truth of the conclusion follows from the characteristics which the present figure has in common with all others of the class, then it will be true for all such figures, and the peculiar characteristics will have nothing to do with the case. The major premise underlying demonstration by means of figures is this: "The present figure is an adequate representative of all figures to which the present proof applies."

The second special form of proof which we shall examine is found in what is known as *mathematical induction.*

PROOF BY MATHEMATICAL INDUCTION.—The following illustration is a typical example of reasoning of the sort just mentioned. "If we take the first *two* consecutive odd numbers, 1 and 3, and add them together the sum is 4, or exactly *twice two;* if we take *three* such numbers, $1 + 3 + 5$, the sum is 9, or exactly *three times three;* if we take *four*, namely $1 + 3 + 5 + 7$, the sum is

16, or exactly *four times four;* or generally, if we take any number of the series, $1 + 3 + 5 + 7 + \ldots$, the sum is equal to the number of terms multiplied by itself. Any one who knows a little algebra can prove that this remarkable law is universally true, as follows: Let n be the number of terms, and assume for the moment that this law is true up to n terms, thus:

$$1+3+5+7+\ldots+(2n-1)=n^2$$

Now add $2n + 1$ to each side of the equation. It follows that:

$$1+3+5+7+\ldots+(2n-1)+(2n+1)=n^2+2n+1$$

But the last quantity, $n^2 + 2n + 1$, is just equal to $(n+1)^2$; so that if the law is true for n terms it is true also for $n+1$ terms. We are enabled to argue from each single case of the law to the next case; but we have already shown that it is true of the first few cases, therefore it must be true of all."[1] If what is true of any case is true of the one following it, it will be true of all cases whatsoever. It sometimes happens that something is true of a great many successive cases without being really general. To quote again from Jevons: "It was at one time believed that if any integral number were multiplied by itself, added to itself, and then added to 41, the result would be a prime number, that is, a number which could not be divided by any other integral number except unity; in symbols, $x^2 + x + 41$ = prime number. This was believed solely on the ground of trial and experience, and it certainly holds

[1] Jevons, *Lessons in Logic*, pp. 220-221.

for a great many values of x.... No reason, however, could be given why it should always be true.... it fails when $x = 40$." "We can perceive no similarity between all prime numbers which assures that because one is represented by a certain formula, also another is; but we do find such similarity between the sums of odd numbers."

Here, as elsewhere, if one or a few cases are adequate representatives of a whole class of cases, what is true of the present case or cases will be true of all, and a universal conclusion can be drawn from a single case. The great difficulty in ordinary inductions is to be sure that the given case is an adequate representative. Ordinarily the facts are very complex and the inspection of a single case is not sufficient to show us the characteristics which an adequate representative of the class should possess. In other words, we do not know what circumstances are relevant. In selecting cases for the application of the several inductive methods, the selection is for the purpose of enabling us to determine what circumstances are relevant.

II. Failure to Prove: Fallacies.—Let us examine the various ways in which proof may be vitiated, the various fallacies to which reasoning is liable. Some of these have been discussed already, but they will be mentioned again here and the other fallacies not previously noted will be examined.

Fallacies of Language.—In the first place we may mistake the meaning of the premises owing to the fact that we have not understood the language in which they are expressed. The *Fallacies of Amphiboly, Accent, Figure of Speech* (see chapter v), are cases in point.

Again, to mistake the general for the specific use of a term, or the concrete for the abstract, or to use a term in one sense in one part of the reasoning and in another sense in another part, would render the conclusion unsound; the *Fallacy of Accident* must be guarded against. (See page 55.)

Once more, we may mistake the collective for the distributive use of a term or vice versa, or we may use a term in one of these senses in one part of the reasoning and in the other sense in another part; this would involve us in a *Fallacy of Composition or of Division.* (See page 57.)

FALLACIES OF ASSUMPTION.—Again, if we use as a premise in our reasoning a proposition which is not established, such as an insufficiently verified inductive inference, our conclusion is not proved. It may be true, or it may not be, but a conclusion is not proved so long as there is a possibility that it may be false. When we use a proposition of this sort to support a conclusion we are said to commit the *Fallacy of Begging the Question*, or, to use the scholastic term, *Petitio Principii.* It is frequent in cases in which one has a thesis to prove and he sees that a certain proposition will enable him to prove it (it may be either premise in a syllogism—or both). He therefore assumes the truth of this proposition on insufficient grounds, sometimes on very slight grounds.

To decline to admit a premise because we see that it necessitates an unwelcome conclusion is to commit this fallacy.

A false categorical proposition, an incorrect hypothetical proposition, or a disjunctive proposition in which the disjunction is not complete, would also, if used as premises, be illustrations of this fallacy. It is not even necessary to use a complete proposition to commit this fallacy; the use of a name or an epithet may lead to fallacious conclusions; the name or epithet does, to be sure, imply a proposition. To argue that this criminal should be punished, because all criminals are a menace to society, begs the question if it has not been shown that this man is a criminal. The epithet is sometimes more dangerous than the implied proposition would be, for we are less likely to notice that something has been assumed when the proposition is only suggested.

One form of the *Petitio Principii* is *Arguing in a Circle*, or *Circulus in Probando.* In this, the premise is simply the desired conclusion stated in other words. To say that man is a conscious being because he has mental states, is to argue in a circle, since to be a conscious being is to have mental states. When the argument is short there is little to be feared from this fallacy; the identity in meaning of the two statements is easily discovered if they come close together; but in a long argument the first may be only vaguely remembered by the time the second is made, and if the reasoning has not hitherto been questioned the fallacy may escape detection. A language like English, which is very rich in synonyms, offers very many occasions for this fallacy.

Another closely related fallacy is that known as the *Fallacy of Many Questions* or sometimes *Double Question.* One of the traditional illustrations is, "Have you left off beating your wife?" Whichever answer is given, Yes or No, seems to admit the truth of the implication. In this fallacy, the question assumes the truth of something which is not proved or admitted, and which may be false. It demands a direct answer, and no direct answer can be given without an apparent admission of the thing assumed.

FORMAL FALLACIES.—There are several fallacies which result from the violation of the principles of syllogistic reasoning. These are usually called the *Formal Fallacies*, because they are said to result from violating the *formal laws* of the syllogism, the laws relating to the number of terms and the distribution of terms in the syllogism. We might include also, *Fallacies of Illicit Conversion and Obversion.* Creighton includes them in *Fallacies of Interpretation.*[2] Using four terms instead of three gives the *Fallacy of Four Terms.* Distributing the major term in the conclusion when it has not been distributed in the premise gives the *Fallacy of Illicit Distribution of the Major Term,* and distributing the minor term in the conclusion when it was not distributed in the premise gives the corresponding *Fallacy of the Minor Term.* Failing to distribute the middle term gives the *Fallacy of Undistributed Middle.* There are also the *Fallacies of Two Negative Premises, and Two Particular Premises.*

This is the way in which violations of the laws of syllogistic reasoning have usually been classified. As we have already seen, these violations can also be dealt with as failures to comply with the principles of the Four Figures. The latter method is less formal and more in accordance with our ordinary habits of thought.

[2] See his *Logic,* chapter xii.

The *Fallacies of Hypothetical Reasoning*, *Denying the Antecedent* and *Affirming the Consequent*, belong in the class just discussed.

THE CONCLUSION MAY NOT FOLLOW FROM THE PREMISES.—There is a form of false reasoning known as the *Non Sequitur*. In this the premises may be clear and true and there may be no fallacies of distribution or of negative premises, but the conclusion does not follow from the premises. It may be true enough and provable on other grounds but it does not belong to the propositions on which it has been based. It was originally called the *Fallacy of False Consequent* and had to do with hypothetical reasoning, but the term *Non Sequitur* is now applied to categorical reasoning in which the conclusion does not follow from the premises. De Morgan's illustration (quoted by Hyslop) is as follows:

> Episcopacy is of Scripture origin.
> The Church of England is the only Episcopal Church in England.
> Therefore, the church established is the church that should be supported.

This fallacy, like the rest, is more likely to pass unnoticed in a long argument than in a short one.

A similar fallacy, sometimes treated as a form of the last, is that of *False Cause*, or *Non Causa pro Causa*, or *Post Hoc ergo Propter Hoc*. This consists in arguing that because one thing has followed another, it is therefore the effect of that other, as if one should argue that because a panic followed the adoption of a certain measure, it was therefore caused by that measure.

MISSING THE POINT.—One more fallacy is to be

noted; that in which the argument is not to the point. The reasoning may with entire correctness prove something but it is not the thing which was to be proved. An opponent is sometimes charged with shifting the ground of debate; that means usually that he is no longer trying to prove the thesis with which he started but something else more or less closely related to it. This is known as the *Fallacy of Ignoratio Elenchi.* Several forms have been distinguished. One of these is the *Ad Hominem* argument; in this, instead of being to the point, the argument is directed against the character or consistency, etc., of the opponent or some other person. When an advocate proves the prisoner's good character and assumes that he has proved his innocence of the crime with which he is charged, he is guilty of this fallacy. When a debater attacks his opponent instead of proving his thesis he commits this fallacy. It is a method of silencing an opponent but not of proving a case. Appeals to authority, to prejudice, to emotion, are all forms of the fallacy of *Ignoratio Elenchi*, as is also the argument in which the victory depends upon the fact that the opponent has not the information necessary to enable him to meet the argument.

When we assume that a proposition is false because the arguments in its support have been discredited, we commit this fallacy. The proposition is only *not proved*, instead of being disproved. A good cause may suffer from bad arguments because of the widespread tendency to commit this fallacy.

In most arguments many of the propositions involved are unexpressed. In such cases it is often difficult to know what fallacy to charge against reasoning which is obviously unsound. Suppose we have the argument, "A classical course is useless because it trains for no profession;" what fallacies might be charged? In the first place we might say that the fallacy was a Non Sequitur; the conclusion does not follow from the grounds which have been stated. It might be replied that there was a further premise understood, namely, "Every college course which does not train for a profession is useless." The fallacy of Non Sequitur would be disposed of, but it would now be possible to charge the fallacy of Begging the Question, in the premise which has been supplied. Or it might be that the premise understood was "Most courses of study which do not train for a profession are useless." That might possibly be true, but even admitting it, the reasoning is not valid, because it violates the principles of syllogistic reasoning.

In cases of doubt the only way of being certain that we have been fair to an absent opponent or have met all replies to our criticisms, is to follow some such procedure as that just illustrated and show that if one criticism can be met another cannot, or that, if the conclusion is to be established, such and such propositions must be shown to be true.

These fallacies are usually discussed in connection with the syllogism, but they may occur in the more complicated forms of reasoning as well. As we have already seen, the syllogism is the typical form of deductive reasoning, but there is much reasoning that is extra-syllogistic: although this might perhaps be put into syllogistic form such an operation is unnecessary if the premises and the steps in the argument are clear.

In more complicated trains of reasoning which involve induction as well as deduction we must make sure not only of the reasoning processes, and of the clear statement of the premises, but also of the soundness of the premises, and of the grounds on which they are based.

EXERCISES

In the following exercises, supply missing premises, state the Figure in which the argument falls, and criticise fully the reasoning, noting the fallacies of every sort:

1. Personal deformity is an affliction of nature; disgrace is not an affliction of nature; personal deformity is not a disgrace.

2. All paper is useful, and all that is useful is a source of comfort to man; therefore, all paper is a source of comfort to man.

3. If Cæsar were a tyrant, he deserved to die; but he was not a tyrant, and therefore did not deserve to die.

4. Every one desires his own good; justice and temperance are everyone's good; hence, every one desires justice and temperance.

5. Some of the inhabitants of the earth are more civilized than others; no savages are more civilized than other races; therefore, no savages are inhabitants of the earth.

6. He must be a Mohammedan, for all Mohammedans hold these opinions.

7. He must be a Christian, for only Christians hold these opinions.

8. All valid syllogisms have three terms; this syllogism has three terms, and is therefore valid.

9. None but despots possess absolute power; the Czar of Russia is a despot; therefore, he possesses absolute power.

10. The right should be enforced by law; the exercise of the suffrage is a right, and should therefore be enforced by law.

11. Nothing is better than wisdom; dry bread is better than nothing; therefore, dry bread is better than wisdom.

12. Every rule has exceptions; this is a rule and there-

fore has exceptions; therefore, there are some rules that have no exceptions.

13. For those who are bent on cultivating their minds by diligent study, the incitement of academical horrors is unnecessary; and it is ineffectual for the idle and such as are indifferent to mental improvement; therefore, the incitement of academical honors is either unnecessary or ineffectual.

14. Suicide is not always to be condemned, for it is but voluntary death, and this has been gladly embraced by many of the greatest heroes of antiquity.

15. Theft is a crime; theft was encouraged by the laws of Sparta; therefore, the laws of Sparta encouraged crime.

16. Nothing but the express train carries the mail, and as the last train was an express, it must have carried the mail.

17. Protective laws should be abolished, for they are injurious if they produce scarcity and useless if they do not.

18. Whosoever intentionally kills another should suffer death; a soldier, therefore who kills his enemy should suffer death.

19. The people of the country are suffering from famine; and as A, B, and C are people of the country, they are therefore suffering from famine.[3]

20. Each of the books in the library is large; hence the library is large.

21. Hunger is a sign of health; therefore, famine which causes hunger is a good thing.

22. Arsenic will kill a man; hence, this medicine will kill you as it contains arsenic.

23. The coat, hat and dress were each in good taste; therefore, the costume as a whole was in good taste.

24. You can always trust to the majority to do what is right in the long run; this man is a member of the majority, and therefore he can be trusted to do what is right in the long run.

25. Eating opium degrades and brutalizes a man; hence

[3] Most of the examples from 1 to 19 were borrowed from Hyslop's *Elements of Logic.* A good many of them belong to the common stock.

DeQuincy and Coleridge were low and degraded creatures.

26. It is wrong to take life of fellow creatures; hence it is wrong to kill a mad dog.

27. Human life will at some time disappear from the earth, for every man must die.

28. America is a Christian country; hence, every American is a Christian.

29. The members of the college are students, teachers, and administrative officers. The members of the football team are members of the college, and hence are students, teachers, and administrative officers.

30. If it is admitted that men who are proficient in engraving are of great service to a community, it must be true that the greater the degree of excellence possessed by the counterfeiter, the better for the government.

31. You must allow that this measure will do untold good to the country—that the whole community will prosper and that our nation will take its place with the foremost. You say you grant all this and still you maintain that it will ruin your particular section. Is not your section a part of the nation, and will it not be benefited as well as the rest of the country?

32. The population of the United States increased 20% between 1890 and 1900; hence, the population of Vermont must have increased at that rate during the same period.

33. Slavery was harmful to the development of the whole country, and hence to the South.

34. Policemen must arrest all persons who block the highways or interfere with traffic. The policeman at this crowded corner does this, and should therefore be arrested.

35. Kant held that all the proofs for the existence of God were fallacious. He was therefore an atheist.

36. At the time of the Galveston flood men worked sixteen hours a day; hence, to talk of an eight-hour day as a necessity for the working classes is absurd.

37. The evidence of the creator is the thing created.

38. Before you stands the vile wretch who has been accused of murder.

39. Why has man one more rib than woman?

40. The candidate is very fond of children, and so no doubt she would be a good kindergarten teacher.

41. This man's arguments are worthless, for he is notoriously dishonest.

42. In answer to the argument that women, as intelligent human beings, are entitled to all the privileges of citizenship, I ask you: Are not women like our sainted mothers, who never held a ballot in their hands, good enough for us?

43. "Woman as well as man should have a part in the world's political affairs; for government is nothing but national housekeeping."

44. "More coffee is consumed in the United States than anywhere else, and America has become the strongest nation."

45. My opponent presents a formidable array of statistics to prove that the country is financially unfit for war; to which I am proud to reply that the old flag has never yet touched the ground.

46. Agassiz did not accept the theory of Evolution; hence I, who know very little of biology, am not justified in accepting it.

47. This man was a good football player, and hence will be a good man to write up the present football situation.

48. A vacuum is impossible, for if there is nothing between two bodies they must be in contact.

49. The government should be in the hands of the Democratic party, for the country could not help prospering under the supervision of the followers of Jefferson.

50. It is indeed an opinion strangely prevailing amongst men, that houses, mountains, rivers, and in a word all sensible objects, have an existence, natural or real, distinct from their being perceived by the understanding. But with how great an assurance and acquiescence soever this principle be entertained in the world, yet whoever shall find it in his heart to call it in question may, if I mistake not, perceive it to involve a manifest contradiction. For, what are the fore-mentioned objects but the things we perceive by sense? and what do we perceive besides our own ideas or sensations? and is it not plainly repugnant that any one of these, or any combination of them, should exist unper-

ceived?—Berkeley, *Principles of Human Knowledge*, Sec. 4.

51. If there were external bodies it is impossible we should ever come to know it; and if there were not we might have the very same reasons to think there were that we have now. Suppose—what no one can deny possible—an intelligence without the help of external bodies, to be affected with the same train of sensation or ideas that you are, imprinted in the same order and with like vividness in his mind. I ask whether that intelligence hath not all the reason to believe in the existence of corporeal substances, represented by his ideas and exciting them in his mind, that you can possibly have for believing the same thing?—Berkeley, *Principles of Human Knowledge*, Sec. 20.

52. In the business of gravitation or mutual attraction, because it appears in many instances, some are straightway for pronouncing it universal; and that it attract and be attracted by every other body is an essential quality inherent in all bodies whatsoever. Whereas, it is evident that the fixed stars have no such tendency towards each other; and so far is that gravitation from being essential to bodies that in some instances a quite contrary principle seems to show itself; as in the perpendicular growth of plants in the elasticity of the air.—Berkeley, *Principles of Human Knowledge*, Sec. 100.

53. "The family, the state, religion and morality are all in danger in this country on account of divorces, according to the speakers at an Episcopal meeting in New York on Sunday. But are things in so bad a way?" In England, "so 'horrible' were the revelations of angry discontent with the married state made by hundreds of the correspondents of a London paper, that it was compelled recently to bring a discussion of the marriage question to an abrupt end."

54. In arguing against the Darwinian Hypothesis, Agassiz is said to have urged the following: "If species do not exist, how can they vary?"

55. Vegetarianism is a healthy diet, for all vegetarians find it so.

56. No educated, much less a scientific person, who is

convinced of the immutable order of things, can nowadays believe in miracles.—Buchner, *Force and Matter.*

57. Either the laws of nature govern, or the eternal reason governs; if both govern together they must be in continual conflict; the government of the latter would render that of the former unnecessary, whilst the action of unalterable laws admits of no personal interference, and can on that account scarcely be called governing. A main point in the proof that the laws of nature are those of reason is, that by thought we are able to deduce other laws of nature from those known to us, so that we find them in experience, and if this does not happen, we naturally conclude that we have formed erroneous conclusions.—Buchner, *Force and Matter.*

58. In all parts of knowledge, rightly so termed, things most general are most strong; thus it must be, inasmuch as the certainty of our persuasion touching particulars dependeth altogether upon the credit of those generalities out of which they grow.—Hooker, *Ecclesiastical Polity,* i, 12.

59. A spirit independent of nature cannot exist; for never has an unprejudiced mind, cultivated by science, perceived its manifestations. . . . How is it possible that the unalterable order in which things move should ever be disturbed without producing an irremediable gap in the world, without delivering us and everything up to an arbitrary power, without reducing all science and every earthly endeavor to a vain and childish effort?—Buchner, *Force and Matter.*

60. Order and progress are two incompatible elements. Progress is accompanied by disorder, by anarchy. For what is progress if not precisely the overturning of a given social order so as to institute a new one?—Draghicesco, *Le probleme de la conscience.*

61. Sunshine is necessary for plants; for vegetable organisms can not increase in size, sending roots into the soil and stems into the air, without the light and heat of the great solar luminary.

62. Nothing is so bad that it cannot be worse.

63. "The canals are not so maintained. They are falling into decay and disuse. The old boats are rotting and

few new boats are built. The business of the canals falls off, and the city of New York, which thirty years ago had 75 per cent. of the foreign trade of the country, now has less than 50 per cent."

64. "Philosophy bakes no bread." Then why waste time upon it?

65. Men have a right to vote. Then where is the justice of depriving criminals of this right?

66. Two and three are even and odd; two and three are five; hence five is even and odd.

67. To inflict capital punishment is to violate one of the commandments in the Decalogue.

68. "The French drink more wine than any other nation and in literature and art they occupy a foremost place."

69. "... it deals with the question exactly when the monstrous tariff is to be tenderly revised by its friends. The answer is 'Never.' The thing cannot be done in prosperous times, because it would disturb business. In a period of depression, it is out of the question, as we then have troubles enough without opening Pandora's box. When affairs are just betwixt and between, neither very good nor very bad, no sagacious Republican would think of meddling with the tariff. Therefore, we say the exact position of the Republicans is, that an unjust tariff is crying out for urgent revision, that they are the only ones who can do the work, and that they will do it just one day after never."

70. "The justice complains bitterly that the court has been obliged to resort to subterfuges in order to employ competent process-servers, the eligible lists providing only worn-out soldiers who lack the essentials of youth, determination, agility and vigor. He is therefore in favor of going back to the discredited system of 'pass examinations' or of re-enacting the *starchless* civil service law."

71. In the domain of physics, to the exploration of which Lord Kelvin has devoted an honored lifetime, he would be a bold man who would cross swords with him. But for dogmatic utterance on biological questions there is no reason to suppose that he is better equipped than any person of average intelligence . . . in the latter (organic nature) scientific thought is 'compelled to accept the idea of creative power.' That transcends the possibilities of

scientific investigation . . . Lord Kelvin, in effect, wipes out by a stroke of the pen the whole position won for us by Darwin. And in so doing, it can hardly be denied that his present position is inconsistent with the principle laid down in his British Association address at Edinburgh in 1871.—Extracts from letters written to the London Times apropos of Lord Kelvin's assertion regarding the limits of science. (Reprinted in *Science.*)

72. The fact is so improbable that extremely good evidence is needed to make us believe it; and this evidence is not good, for how can you trust people who believe in such absurdities?

73. The axioms of mathematics and the fundamental moral principles are inborn; for they are accepted by everybody. Moreover no reasonable being can deny them when he understands what they mean.

74. I may doubt everything except that I think. I think, therefore I exist.

75. A parliamentary government is sure to fail in the long run; a battle may be won by a poor general but never by a debating society.

76. Can anything be more ludicrous than first to build all our certainty of the assistance of the Holy Ghost upon the certainty of tradition and then afterwards to make the certainty of tradition to rely upon the assistance of the Holy Ghost.—Tillotson, *Rules of Faith.*

77. If men are not likely to be influenced in the performance of a known duty by taking an oath to perform it, the oaths commonly administered are superfluous; if they are likely to be so influenced, every one should be made to take an oath to behave rightly throughout his life; but one or the other of these must be the case; therefore either the oaths commonly administered are superfluous or every man should be made to take an oath to behave rightly throughout his life.[4]

78. Few treatises of science convey important truths, without any intermixture of error, in a perspicuous and interesting form; and therefore, though a treatise would deserve much attention which should possess such excellence,

[4] The exercises from 77 to 90 are from Whately's *Elements of Logic.*

it is plain that few treatises of science do deserve much attention.

79. No one who lives with another on terms of confidence is justified, on any pretense, in killing him; Brutus lived on terms of confidence with Cæsar; therefore, he was not justified, on the pretense he pleaded, in killing him.

80. He that destroys a man who usurps despotic power in a free country deserves well of his countrymen; Brutus destroyed Cæsar, who usurped despotic power in Rome; therefore, he deserved well of the Romans.

81. Nothing which is of less frequent occurrence than the falsity of testimony can be fairly established by testimony; any extraordinary and unusual fact is a thing of less frequent occurrence than the falsity of testimony (that being very common); therefore, no extraordinary and unusual fact can be fairly established by testimony.

82. Testimony is a kind of evidence which is very likely to be false; the evidence on which most men believe that there are pyramids in Egypt is testimony; therefore, the evidence on which most men believe that there are pyramids in Egypt is very likely to be false.

83. He who cannot possibly act otherwise than he does, has neither merit nor demerit in his action; a liberal and benevolent man cannot possibly act otherwise than he does in relieving the poor; therefore, such a man has neither merit nor demerit in his action.

84. The religion of the ancient Greeks and Romans was extravagant fables and groundless superstitions, credited by the vulgar and weak, and maintained by the more enlightened, from selfish or political views; the same was clearly the case with the religion of the Egyptians; the same may be said of the Brahminical worship of India, and the religion of Fo, professed by the Chinese; the same, of the mythological systems of the Peruvians, of the stern and bloody rites of the Mexicans, and those of the Britons and Saxons; hence we may conclude that all systems of religion, however varied in circumstances, agree in being superstitions kept up among the vulgar, from interested or political views of the more enlightened classes.

85. What happens every day is not improbable; some things against which the chances are many thousands to

one, happen every day; therefore, some things against which the chances are many thousands to one are not improbable.

86. The principles of justice are variable; the appointments of nature are invariable; therefore, the principles of justice are not appointments of nature.

87. Of two evils, the less is to be preferred; occasional turbulence, therefore, being a less evil than rigid despotism, is to be preferred to it.

88. No evil should be allowed that good may come of it; all punishment is an evil; therefore, no punishment should be allowed that good may come of it.

89. Repentance is a good thing; wicked men abound in repentance; therefore, wicked men abound in what is good.

90. If the exhibition of criminals, publicly executed, tends to heighten in others the dread of undergoing the same fate, it may be expected that those soldiers who have seen the most service, should have the most dread of death in battle; but the reverse of this is the case; therefore, the former is not to be believed.

91. Why does a ball, when dropped from the masthead of a ship in full sail, fall not exactly at the foot of the mast but nearer the stern of the vessel?—Davis, *Logic.*

92. "The impious, whoever he may be, ought not to go unpunished. For do not men regard Zeus as the best and most righteous of the gods? And even they admit that he bound his father because he wickedly devoured his sons."—Plato, *Euthyphro.*

93. The soul is unchangeable; the unchangeable is simple; the simple is indissoluble; the indissoluble is indestructible; therefore, the soul is immortal. See Plato, *Phaedo.*

PART II

SUPPLEMENTARY METHODS

CHAPTER I.

STATISTICS

The value of any conclusion depends largely upon the soundness of the premises from which it is drawn; a great many of these premises, as we have seen, are inductions from particular facts. When these inductive inferences have been tested by one or another of the "Inductive Methods" they can be regarded as trustworthy; but the successful application of the methods presupposes a fairly complete analysis of the phenomena under investigation, for it is by this analysis that we determine the circumstances in which the phenomenon occurs. If we cannot determine the circumstances it is obvious that we cannot apply the Methods. It might seem to follow from this that if the circumstances in which a phenomenon occurs are so complex as to defy analysis, or if the phenomenon itself cannot be separated into its elements, it would be impossible to make any reliable generalizations regarding the relations of the phenomenon in question. Or if we were quite unable to surmise which of a multitude of circumstances was significant, or to isolate any of them by means of the Methods, we could not decide which was causally related to the phenomenon and which was not. There are many fields in which analysis is possible in only a slight degree; social phenomena and phenomena of the weather are cases in point. A moment's reflection will show the difficulty of applying the Method of Agreement, for instance, in the study of the weather, or of the death rate. The phenomena are so exceedingly complex that anything approaching a complete statement of their elements is quite out of the question. The fallibility of most popular generalizations in these fields is evidence of the difficulty of dealing with such facts. Must we be content then simply to guess at the relations of such phenomena, with the slight assistance which is to be gained from so precarious a method as that of Simple Enumeration? In instances of this sort another method, a method which is closely related to the method of Simple Enumeration, becomes important: it is the Method of Statistics. In statistics we have an *exact* enumeration of cases. If a small number of cases does not enable us to detect the causal relations of a phenomenon, it sometimes happens that a large number, accurately counted, and taken from a field widely extended in time and space, will lead to a solution of the problem.

But how can the counting of cases aid in the discovery of a causal relation? It does so by showing the relative frequency of the phenomenon, its frequency as compared with some particular circumstance or circumstances. If we noted only the phenomenon itself, knowledge of its frequency would be of little use. But if, in a large number of cases, taken from a wide field, we can find some other phenomenon correlated with the one we are investigating, then we have ground for a conclusion. We proceed upon the principle that if the concurrence of two phenomena is merely a coincidence, the frequency of one should make no difference in the occurrence of the other, and conversely that *if there is a correlation, there must be some causal relation.* If the frequency of two phenomena is the same, or if variations in the frequency of one correspond to variations in the frequency of the other, or if any change in the quantity or quality of one corresponds to changes in the frequency of the other, we are usually justified in inferring that something more than a coincidence is present.

Correlation may be either positive or negative; in positive correlation the presence of a phenomenon A would mean the presence of the phenomenon B in a certain proportion of cases or in a certain amount, and so on; in negative correlation the presence of A would of course mean the absence of B in a certain proportion of cases, and so on. The relation between illiteracy and crime would be an instance of the former and the relation between vaccination and smallpox would illustrate the latter. The total absence of correlation might perhaps be represented by the relation between the weather and the day of the month. Correlation can be measured mathematically; that is to say, it is possible to determine just what degree of correlation there is between two phenomena. The number which expresses this is called the coefficient of correlation. Complete positive correlation would be expressed by + 100, complete negative correlation by - 100, and absence of correlation by 0; A coefficient of + 63 between two phenomena would mean that one was present in 63 per cent. of the cases in which the other was present, or that the amount, or amount of increase, and so on, of one, was 63 per cent. of that of the other.[1]

Sometimes [2] a correlation would not prove a direct causal relation; the fact that the mortality among men is higher than that among women and bears a certain numerical relation to it, or the fact that about 106 boys are born for every 100 girls, are examples of this. But there is nothing peculiar to statistics in this. The same thing appeared in connection with the method of Agreement; here, as there, the concurrence of two phenomena may mean that both are connected with some one underlying set of complex or undiscovered conditions. And even frequent concurrence may be accidental, though a thoroughgoing investigation would eliminate one which was entirely accidental.

And even if the cause of a phenomenon is not discovered in this way, it may be that its frequency is a matter of interest or of practical importance; for its frequency may itself be a factor in determining our conduct; the number of passengers stopping at a given railway station or the comparative number of boys and girls in a city may be worth knowing, even if not understood. The census reports contain a multitude of facts of this kind; eventually the causal relations of many of them may be discovered.

A statistical record like any other enables us to cor-

1 For further discussion, see Thorndike, *Mental and Social Measurements;* Bowley, *Elements of Statistics;* Pearson, *The Chances of Death.*

2 It will be noted that the principles here employed are related to those used in the *Methods of Agreement* and of *Concomitant Variations,* but here analysis may be very incomplete; frequency instead of quantity is considered, and principles of the method of Difference may be employed in conjunction with the others.

rect mistakes of memory, and so on; as, for example, in matters such as the increase or decrease of crimes, or the decreasing or increasing coldness of winters.

Usually the cause discovered by means of statistics is only a *part of* the cause of the phenomenon; the phenomenon is the result of a number of circumstances working together (Composition of Causes), and not always of the same circumstances (Plurality of Causes).

We may know already that a certain circumstance is causally related to a certain phenomenon but that it is sometimes present without the latter (through the agency of a Counteracting Cause or the absence of necessary supplementing circumstances). In such a case statistics enable us to discover how frequently, in what proportion of instances, the phenomenon will be present along with that circumstance.

It must be remembered that the statistical method means more than the mere collection of cases. "With the collection of statistical data, only the first step has been taken. The statistics in that condition are only raw material showing nothing. They are not an instrument of investigation any more than a kiln of bricks is a monument of architecture. They need to be arranged, classified, tabulated, and brought into connection with other statistics by the statistician. Then only do they become an instrument of investigation, just as a tool is nothing more than a mass of wood or metal, except in the hands of a skilled workman."[3]

The processes used in statistical investigations differ widely, but the following are generally given in discussions of the subject: (1) The Collection of Material,

[3] Mayo-Smith, *Statistics and Sociology*, p. 18.

(2) its Tabulation, (3) the Summary, and (4) a Critical Examination of its results. "In collection and tabulation common sense is the chief requisite, and experience the chief teacher; no more than a knowledge of the simplest arithmetic is necessary for the actual processes; but since . . . all parts of an investigation are interdependent, it is expedient to understand the whole before attempting to carry out a part. For summarizing, it is well to have acquaintance with the various algebraic averages, and with enough geometry for the interpretation of simple curves, though all the operations can be performed without the use of algebraic symbols."[4]

The collection of statistics is carried out by various methods, some of them very technical; we can note only a few general principles here. In the first place the data should be collected over a wide field. Just as in the non-statistical application of the inductive methods, it is necessary to collect data over a field wide enough to insure us against mistaking a coincidence for a cause or over-emphasizing the importance of one out of a number of coöperating causes, or regarding as the sole cause one which is *only* one of a number of different causes capable of bringing about the phenomenon.

One danger to be guarded against arises from the failure of different observers to use terms in exactly the same sense. If poverty means in some cases inability to obtain luxuries and in others, positive want, we can make little use of statistics of poverty.

Again, in many statistical investigations, the data are obtained by means of questions addressed to a great

[4] Bowley, *Elements of Statistics*, p. 17.

many individuals; these questions should be so worded as to minimize as far as possible the tendencies to careless or biased observation, faulty memory, prejudice, dishonesty, and imperfect description of the facts.

Tabulation involves classification, and the scheme of tabulation should be determined by the purposes of the investigation—the problems which it is intended to solve. "In general, the scheme of investigation requires knowledge of certain groups; and the totals resulting from tabulation should show the number of items in these, so that after tabulation, instead of the chaotic mass of infinitely varying items, we have a definite general outline of the whole group in question."

The totals and averages must be so presented as to give a true impression to an inquirer. The subject of averages will be discussed more fully in a later chapter.

In the summary, the aim is to present the results in the clearest, most comprehensive, and most suggestive way. The use of averages, and representation by charts and diagrams, are important here. In correlating the results great care is needed to avoid wrong interpretations. An increase in the number of arrests might be causally related to increasing severity in the enforcement of law and not to an increase in crime.

A critical examination of the results is possible only when the sources of the data, the methods of their tabulation, and the mode of summarizing and drawing conclusions, are fully described.

There are, of course, many cases in which the use of statistics would be unnecessary: "In order to prove the relation between savagery and fetichism it is not necessary for us to have statistics either of economic condition or of religious confession. The fact stands out of itself simply by the consensus of observation of travelers and historians."[5]

Where the law of the data is already known the frequency of their occurrence is of no further interest to science. The number of times an acid has combined with a base to form a salt is of no importance to the chemist. If we know the laws and the circumstances, the frequency of the event and the times of its occurrence can easily be determined. "There was some interest in counting how many eclipses of the moon and sun took place every year, so long as they occurred unexpectedly and inexplicably; since the rule has been found according to which they occur and can be calculated for centuries past and to come, that interest has vanished. But we still count how many thunderstorms and hailstorms occur at a given place or within a given district, how many persons die, and how many bushels of fruit a given area produces, because we are not in a position to calculate these events from their conditions."[6]

In other cases the method of statistics may be inapplicable. "It is difficult to express the relation between economic condition or religious feeling and æsthetic development in a civilized state, because music, painting, and sculpture cannot in any way be measured statistically. This is a question of quality and not in any sense of quantity."[7]

[5] Mayo-Smith, *Statistics and Sociology*, p. 9.
[6] Sigwart, *Logic*, Part III, chap. iv, 3.
[7] Mayo-Smith, *loc. cit.* See, however, page 210 on the ways of applying exact methods in investigating such phenomena.

The use of statistics is often severely criticised and there is much popular distrust of the results attained by their employment. There are, of course, many difficulties to be met, and many conclusions based upon statistics may be false. They are liable to most of the errors which occur in connection with the handling of individual facts. The original observations may have been faulty; in so far as memory was employed, further errors may have entered; ignorance, prejudice or inaccurate statements may have vitiated whatever testimony was employed; the records may have been faulty or mistakes may have been made in copying; the facts observed may not have been representative; in comparing different groups and in noting correlations we may mistake a mere coincidence for causal relation. If all the precautions which are employed in a scientific examination of individual facts are made use of here, statistics may furnish a perfectly valid basis for inference. One practical difficulty is the unfamiliarity of the average reader with the use of statistics and his consequent inability to criticise them, and another is the frequent failure on the part of the investigator to furnish data for criticism.[8]

[8] Interesting illustrations of the use of statistics are easy to find. The field of vital statistics is a good one for this purpose. One very interesting study is Dr. Allyn A. Young's *A Discussion of Age Statistics, Bulletin* 13 *of the Bureau of the Census.*

CHAPTER II

AVERAGES

The Arithmetical Average.—In statistical investigations and in all others in which quantitative data are employed, the use of averages is often very important. An average is a single quantity which represents two or more other quantities. There are several kinds of averages; that with which we are most familiar is the Arithmetical Average. It is obtained by adding together the various quantities to be averaged and dividing their sum by the number of quantities. The weights of the members of a college football team were respectively, 175, 195, 187, 183, 230, 187, 169, 147, 159, 178 and 185. The average was 181 4-11. The average is less cumbersome than the whole series of quantities or their sum. The greater the number or size of the quantities the more important does the average become.

The average tells us nothing about the individual cases. In this example the average is not the same as any single one of the quantities averaged. To take another case: the death rate of a city gives no information regarding the death rate of any given ward, nor the number of deaths in any given thousand of the population. An average simply serves as a means for representing the whole series of quantities and for comparing it with other series. It gives no information regarding the homogeneity of the group: 180 is the average of 179 and 181 and also of 359 and 1.

There are many cases in which the simple form of the

arithmetical average or mean is inadequate; sometimes a modified form of it can be used.

The "Weighted" Average.—Suppose we know that of six *groups* of men the average weights are respectively 180, 148, 172, 164, 156 and 152 pounds. The average is 162. Can we say that this average satisfactorily represents the whole series of groups? That will depend upon the circumstances. If the groups were of approximately the same size it might be sufficient, but if in the first group there were 10; in the second, 200; in the third, 50; in the fourth, 20; in the fifth, 100; and in the sixth, 150, our average will be a very imperfect representative of the groups. If, on the other hand, we multiply each of the averages in the series by the number of individuals in the group which it represents and divide the sum of these products by the total number of individuals, we get the average 154 6-53, which is much more accurate than that first given.

$$\frac{180 \text{x} 10 + 148 \text{x} 200 + 172 \text{x} 50 + 164 \text{x} 20 + 156 \text{x} 100 + 152 \text{x} 150}{10 + 200 + 50 + 20 + 100 + 150} = 154\tfrac{6}{53}$$

This is an illustration of what is known as a weighted average; it is a special form of arithmetical average. Where the groups represented by a series of averages vary greatly in size we have conditions which call for "weighting the averages." "The classical and most useful application of weights is the formation of an index-number for the change of prices by fitting suitable weights to the changes measured in the prices of various commodities. It is required to find the change in the value of gold when measured by the prices of other commodities. Suppose that we are given that prices of certain commodities between two years were in the following ratios:

	Wheat	Silver	Meat	Sugar	Cotton
First year	100	100	100	100	100
Second year	77	60	90	40	85

The simplest way to estimate for the general fall in price is to take the simple average of the numbers in the second year, *viz.*, 70.4, and say that the general prices in the second year were 70.4 : 100 [1] when expressed in commodities. But it is at once clear that we can not allow the commodities given to have equal influences on the result; wheat is of greater importance than sugar and meat than silver; and again we have taken arbitrarily three items to represent food and one for clothing; we need some means of deciding relative importance. Suppose we decide that wheat, cotton, meat and sugar are respectively 7, 4, 3, times and twice as important as silver, we should get the following table:

Commodity	Relative prices in second year	Weight Assigned	Product
Wheat	77	7	539
Silver	60	1	60
Meat..............	90	3	270
Sugar	40	2	80
Cotton	85	4	340
	352	17	1289

Weighted average is $\frac{1289}{17} = 75.8$

Unweighted average is $\frac{352}{5} = 70.4$ [2]

[1] That is, prices of commodities have fallen in this rate or the value of gold has increased correspondingly.

[2] Bowley, *Elements of Statistics*, pp. 111-112.

It is not always easy to tell what weights should be assigned, but considerable variation is possible without much modification of the result.

The Mode.—Another sort of average which is often of great importance is what is known as the Mode. It is that quantity which occurs with the greatest frequency. It is what we frequently have in mind when we speak of the average man, the average student, etc. If in a class of students, 10 receive the grade A; 20, the grade B; 50, the grade C; 100, the grade D; and 25, the grade F, the mode is D. The mode very often represents the type more accurately than does the average. It gives us no information about any one individual, but it does indicate the sort of individual which occurs more frequently than any other sort. There might well be two or more modes in a given series of quantities. If a class were made up of very bright and very dull students, the numbers receiving the various grades might be A, 25; B, 50; C, 20; D, 100; and F, 25. The two modes are at B and D.

The mode is not always easy to determine. In these examples the grade B, for instance, means a range between the grade which is just high enough to escape C and that which is the smallest fraction short of A. It might well be that of the 50 who were in C, 35 were in the lower half of the group, while of those that were in D 80 were in the upper half of the group, so that the mode was really in a group which might be indicated by the expression D +, C −. The degree of accuracy required in the results would determine the degree of exactness with which we should state the mode.

The mode is often most useful. "The mode rather than the average in chest-measurements is the number most suitable for the ready-made clothier. For providing a post-office or a store, the mode in postal-orders or prices of tea needs to be known rather than any other average. Even the favorite coin in a collection may show the spirit of the congregation better than the arithmetic average of their contributions." [3]

If the series under consideration is very irregular it may be quite impossible to apply the mode.

The mode has this advantage over the arithmetical average: it is uninfluenced by extreme cases. In the illustration on page 198, the average weight of the players would be considerably changed if a player weighing 180 pounds were substituted for the one weighing 230; let us see what would happen in the case of the mode. More of the quantities fall between 180 and 189 than within any other equal range. This range, 180-189, then, will be the mode. The substitution of the lighter player does not modify the mode. Where the number of quantities is so small as in this illustration, the individual quantities are often mentioned, but where that is not the case, the mode is often useful either as a supplement to the arithmetical average or as a subsitute for it.

The Median.—Another kind of average useful in many cases is the Median. The Median is the middle quantity in a series. The weights of the players, in the order of their magnitude, were 147, 159, 169, 175, 178, 183, 185, 187, 187, 195, 230. The median is 183. There are just as many items above it as there are below. The median, like the mode, is unaffected by extreme cases. "The existence of any number of millionaires has no more effect on the median income than of

[3] Bowley, *Elements of Statistics*, p. 123.

an equal number of other persons whose incomes are above the median."[4] The median is very easy to find, since it is only necessary to arrange the items in the order of their magnitude and find which occupies the middle position. If there is an even number of items the median lies between the two middle ones. Even if our information regarding the items is incomplete it is often possible to find the median with a fair degree of accuracy. "It may be that in the 'wage census' 100,000 persons whose wages were far below the average did not come into the returns at all, and it is very difficult to estimate their effect on the arithmetical average, for want of information as to their earnings; but to find the median exactly we need only know their number, not their earnings; and if we can assign a maximum for their number, we still can place the median within narrow limits."

One great advantage in the use of the median is to be found in the fact that it can be employed in dealing with quantities for which no accurate measurements can be obtained. This is especially important in dealing with psychological phenomena. We may be able to say that A has a better memory than B without being able to measure either, or to state the exact amount in which A is superior to B. The members of a class of any size might be arranged in the order of their excellence in any quality whatever and the median found as in the case of numbers. Francis Galton, in his *Natural*

4 "The magnitudes one-quarter and three-quarters up the series are called the *quartels;* those one, two, ... nine-tenths of the way up are the *deciles;* those one, two, ... ninety-nine hundredths up are the *percentiles.*" Bowley, *Elements of Statistics,* p. 124.

Inheritance and in other works, developed and applied this type of average with great effectiveness.

The median may be a very imperfect representative of the type. If, in a group of 100 men, the weights of 50 were between 190 and 210 pounds, while the others ranged between 130 and 150 pounds, the median would be 170. "The median is then chiefly useful when we are dealing with a series of objects of which the main part lie fairly close together; a few extremes do not affect it."[5]

The Geometrical Average.—Another kind of average useful in certain cases is the Geometrical average. It is related to the arithmetical average somewhat as compound is to simple interest. The population of Great Britain and Ireland increased from 12 millions in 1789 to 38 millions in 1890. Obviously it would be unsafe to say that the average increase was the total increase, 26 millions, divided by the number of years. We should expect the annual increase to be greater as the population became larger, and, other things being equal, the two would vary together. When we have only two quantities to deal with, the geometrical average is easily found. In such a case it is the mean proportional. The geometrical average of 4 and 16 is 8. The geometrical average of 5 and 9 is the square root of 45, or 3 into the square root of 5. If we were dealing with three quantities the geometrical average would be the cube root of their product; if with five, it would be the fifth root of their product; the general formula for n quantities is the nth root of $a_1\ a_2 \ldots a_n$. With large or numerous quantities logarithms should be used. The name *logarithmic mean* is sometimes employed for this

5 Bowley, *Elements of Statistics*, p. 126.

kind of average. The geometrical or logarithmic mean is a quantity which can be substituted for each of the quantities when they are multiplied together and give the same product, whereas the arithmetical mean is one which can be substituted for each of them when they are added together. "Which mean we should choose is simply a question of which we believe will best represent the facts. If the growth of cities depended altogether upon the birth of children within their boundaries, we should naturally choose the geometrical mean, for the larger the city (other things being equal) the more children will be born in it. If, on the other hand, the population of a city, like that of a prison or hospital, were made up altogether of certain kinds of people who were sent there from without, there would (?) be no reason why a large city should gain more inhabitants than a small one; and the more appropriate average would be the arithmetical. With most cities the natural rate of growth is only partly geometrical and partly arithmetical; so that neither a series of means of the one sort nor a series of the other would give a wholly satisfactory representation of the mean growth from year to year between one census and another. If in any case or set of cases we have reason to believe that the true mean lies somewhere between the arithmetical and the geometrical, and if we wish to represent the facts as accurately as they can be represented by any mean, we must take a mean that does lie between the two."[6]

Measuring Deviations from an Average.—It is often important when using averages to know something about the closeness with which the several quantities

[6] Aikins, *The Principles of Logic*, p. 315.

approximate the average. Suppose for example, that we had a number of different measurements of a given quantity, say the distance between two points: if there was little variation among the measurements we should usually regard their average as a fairly accurate representation of the real quantity; but if the variation were very great we might have little or no confidence in the average. We shall need some way of indicating the amount of divergence within the group, or, in other words, the closeness with which the several quantities were grouped about the average.

1. One simple way of doing this is to take the average of the *deviations* from the mean or average. Eight is the average of 5, 6, 11, and 10, and also of 1, 2, 15, and 14. The deviations in the first series are 3, 2, 3, and 2. The average of these deviations is $^{10}/_4$ or $2\frac{1}{4}$. The deviations in the second series are 7, 6, 7 and 6; the *average deviation* is $6\frac{1}{2}$.

(The deviations are technically known as "errors," and their average as the *Average Error.*)

The smaller the average error the more closely are the quantities grouped about the average; and the more closely they are grouped about the average the more homogeneous is the group.

2. Another kind of average frequently employed in this connection is the *Median* or *Probable Error* (P. E.). Arrange the errors in the order of their magnitude; the Median of these will be the so-called Probable Error, or the quantity within which half of the errors fall. Thus, if we have the quantities, 1, 3, 6, 8, 9, 12, 13, 15, 16, 17, the average will be 10. The errors will be 9, 7, 4, 2, 1, 2, 3, 5, 6, 7. Arranging these in the order of their magnitude we have 1, 2, 2, 3, 4, 5, 6, 7, 7,

9. The median will fall between 4 and 5, *i. e.*, $4\frac{1}{2}$. In other words, $4\frac{1}{2}$ is the quantity below which half the errors fall and above which we will find the other half. Assuming that our data are representative of the class of facts for which they stand, any new number standing for things in the same class is as likely to be within $4\frac{1}{2}$ of the average as it is to be beyond it. Exactly half of the quantities already determined lie within that range (in the example, between the numbers $5\frac{1}{2}$ and $14\frac{1}{2}$), and those already determined are, according to our supposition, selected from a wide enough field to be regarded as representative of the whole. An average of 10 with a probable error of 4.5 means a series of quantities in which there is a wide range of variation. An average of 100 and a P. E. of .1 would indicate a very homogeneous group. In the case of measurements it would mean that there was a close agreement among the different measurements and that the average was therefore a fairly accurate approximation to the true measurement (providing, of course, that constant errors had been eliminated).

"An approximation to the probable error for a given series of observations is obtained by arranging all the observations in order of magnitude; marking the magnitude, say *a*, above which 25 per cent. of the observations lie, and the magnitude, say *b*, below which 25 per cent. lie. Half the difference between *a* and *b* is the probable error. A useful way of illustrating this is to say that if one observation is chosen at random out of a group, it is as likely as not that it will not lie further from the average than the probable error." [7]

[7] Bowley, *Elements of Statistics*, p. 282.

Measurement of Phenomena.—In the more advanced stages of most sciences the exact measurement of phenomena becomes more and more important. To determine the relations of a phenomenon it is not only necessary to know when it happens and what its accompaniments are, but also how much of it is correlated with given amounts of other phenomena. This is evident in the employment of the methods of Concomitant Variations, for this method deals with cases in which the quantity of the phenomenon varies. In the method of Residues also, quantitative measurements are of great importance; indeed, they are usually necessary. We observe how much of a given phenomenon is due to one cause, how much to a second, and so on; the remainder is due to something else not previously known to be a cause, etc.

The physical sciences are very largely quantitative, and more recently biology has come to employ the methods of exact measurements in many of its investigations.[8] Measurement usually means the employment of instruments. Measurements of magnitudes by the unassisted eye are exceedingly inexact, and measurements of degree of quality are even more so. White marble painted in a picture representing an architectural view by moonlight seems to be of about the same degree of brightness as the actual moonlit marble would be, but Helmholz has calculated that it is from ten to twenty thousand times as bright.[9]

To make measurements it is necessary to fix units in

[8] The name of Karl Pearson is most closely associated with "Biometry."

[9] James, *Psychology, Briefer Course*, p. 155.

terms of which the magnitudes are to be expressed. These are usually determined arbitrarily. Units, standards, and instruments of measurement vary with the phenomena to be measured and can not be discussed further here.[10]

Many errors may occur in making measurements, and although it is often possible to eliminate some of them, in the vast majority of instances the measurement is almost certainly inexact. Repeated measurements are very seldom in exact agreement. If phenomena were broken up into units of uniform magnitude there would be less difficulty, but most phenomena are continuous. Time is not broken up into minutes, and with the most exact instruments it is impossible to say when a minute has passed. It can be determined within millionths of a second but not with absolute exactness. For most practical purposes rough measurements are sufficient; thus, for the train dispatcher it may be enough to determine the time to a second, but for astronomical calculations the smallest possible error may be of serious importance. Most measurements are only approximately true; the problem is to make the approximation as close as possible.

There are various conditions to be observed and various methods which can be used in attempting to get exact measurements, but they are too technical to be included here.[11] Constant errors, such as the personal error,[12] can often be determined and allowance can be made for them. But after all such allowances

[10] See Jevons, *Principles of Science,* chap. xiv.
[11] See Jevons, *Principles of Science,* chap. xiii.
[12] See page 23.

have been made and after all the means for avoiding and minimizing error have been employed, there yet remains a margin of uncertainty. In such cases it is possible to obtain a close approximation to the true measurement by taking a number of measurements and striking an average. After constant errors have been eliminated any given measurement is as likely to be too great as it is to be too small; hence, in a large number of measurements there will probably be as many of those which exceed the true magnitude as there are of those which fall short of it. If the number of measurements is small this is more doubtful, but if a great many measurements have been made, we can rely upon the average with safety. The average of all these measurements is the closest approximation which we can get. Different kinds of average are used according to circumstances. The closeness with which the several measurements are grouped about the average will be indicated here, as in all cases of the use of average, by the size of the error. If the error is small, the measurement is reliable, if large, more doubtful.

The Comparison of Quantities which Cannot be Measured.—In the study of many phenomena the problem of quantitative comparison is made very difficult by our inability to find an exact quantitative equivalent for the phenomena. "Many mental phenomena elude altogether direct measurement in terms of amount. How many thefts equal in wickedness a murder? If the piety of John Wesley is 100, how much is the piety of St. Augustine? How much more ability as a dramatist had Shakespeare than Middleton? What per cent. must be added to the political ability of the Jewish race to make it equal to the Irish race? . . . Nevertheless, such phenomena can be measured and subjected to quantitative treatment." [13]

The method to be employed in such cases, as Professor Thorndike goes on to show, is to arrange the individuals (or other unmeasurable data) according to their rank. We may not be able to say how much more eminent A is than B, but if we can say that A is in the first rank, whereas B is in the tenth, we have a true basis of comparison. We cannot measure directly the intelligence of students in a class, but we may be able to say that one is in the first group, whereas another is in the fourth. Thus, with any number, it would be possible to give each his proper place in the group. This method can be applied to any trait whatever. The great difficulty is in making sure that the ranking is correct. Single observations and individual judgments are subject to the same errors here as in all other cases of observation.

EXERCISES.

1. What sort of average should be employed in determining the standard size of an article to be manufactured in large quantities—say window shades?

2. What sort of average should be employed in getting a number to represent the value of articles in a large and varied invoice of merchandise?

3. If a college had 400 students in 1880 and 1000 students in 1905, how many did it have in the year which falls

[13] Thorndike, *An Introduction to the Theory of Mental and Social Measurements,* p. 18. This book is an exposition of the methods of measuring individuals, groups, variability of performances, etc., including an exposition of the necessary modes of presenting the facts, making calculations, and so on.

half way between, provided that the rate of increase was constant?

4. What averages might be employed and which would be preferable in comparing the stature of soldiers in the French army with those in the American army? In comparing the standing of successive classes in college? In comparing the salaries of members of the faculty in two universities? In comparing the rate of growth of a large university and a small college?

5. How would you indicate the degree of closeness with which a series of quantities approached their average?

6. What is the difference between "Average Error" and "Probable Error?"

CHAPTER III

PROBABILITY

The conclusions at which we arrive by the assistance of statistical methods and the employment of averages often fall far short of the certainty attaching to scientific laws. The conditions required for establishing a scientific law are not fully present, and consequently many of such conclusions, if not all of them, lack complete verification. It does not follow, however, that these are valueless. As a matter of fact, most of the generalizations which we use in everyday life are incompletely verified; they are extremely valuable as instruments of knowledge and practice; indeed, in the absence of scientific laws, they are indispensable. So long as their provisional character is remembered, there is no serious danger in using them.

A generalization of this character is said to be probable or to possess some degree of probability. Probability belongs also to particular propositions. What do we mean by probability,—by saying that a statement is probably true,—that an event will probably happen? As we use the term ordinarily, it means that we believe we have a right to accept a statement or expect an event, without feeling perfectly certain of it. This attitude, when it has any justification, is based upon the belief that the grounds for accepting the statement are stronger than those for rejecting it. It may be that we know of no positive reasons against it, but do not regard the reasons in its favor as conclusive; or it may be that there are positive reasons against it, but that those in its favor are stronger or more numerous. These reasons or grounds may be of various kinds. There may be many things pointing toward the occurrence of such an event; as, for example, in the statement that life will probably at some time cease upon the earth. Or conditions at the present time may be similar to those in which the event has happened before; the outcome of an examination of instances according to the principles of the method of Agreement gives a result which is usually only probable. In all these cases it is impossible not to feel that a great deal of vagueness attaches to our statement that anything is probable. We are not able to say how probable it is. There is such a thing, however, as mathematical or quantitative probability. It is based upon the comparative number of times an event or connection of events has occurred. If a given circumstance A has been observed 1000 times, and if, in 700 cases of its occurrence, a phenomenon B has also been present, we have definite grounds for inferring that A will probably be accompanied by B again. Every time A and B have occurred together in the past is an argument in favor of their occurring together in the future, and every time A has occurred without B is an argument against this connection; if the cases of the latter sort are many in comparison with those of the former, we say that the connection in the future is improbable. In the case just mentioned we should express the degree of probability by the fraction $7/10$. Now in dealing with the matter in this quantitative way, the term "probability" has a meaning which is somewhat different from that in which we ordinarily use it. It would mean, in the present case, that in the future we should have a right to expect B along with A in seven cases out of ten. It means nothing with regard to the next case; we have no more reason for expecting one outcome than the other; our information has value only for the long run; we have no right to expect that in the next ten cases B will be present seven or any other particular number of times; but in the long run we may expect this proportion to hold and the longer the run the closer is the approximation which we may expect.

Compare this with one of the former illustrations, "Life will probably cease upon the earth." That does not mean that in a large number of cases of the sort before us life would cease in most of them; we are here dealing with the particular case and all our arguments apply to it. In quantitative probability we know nothing of the circumstances of the particular case; it is simply one of a certain group, and certain members of this group behave in one way, whereas others behave in a different way, and we cannot determine the circumstances of their behavior in either case. The fraction expressing the degree of probability tells us that, in the past, the phenomenon has appeared in connection with a certain circumstance in such and such a proportion of cases and that we may, unless there are reasons to the contrary, expect this proportion to hold in the future. (We proceed here as elsewhere, upon the assumption that the future will be like the past and that any set of phenomena will behave in the future as it has in the past, in the absence of any new and disturbing factor.)

Probability, in this connection, does not necessarily mean favorable odds. The event may have occurred in only one-tenth of the cases; its probability will then be $1/10$. If it has occurred in one-half the number of cases, the probability will be $1/2$, etc.

The calculations of insurance companies are based upon data showing the number of deaths per year for individuals of various ages, and so on. The great value of vital statistics and of statistics of many other sorts is to enable us to determine the probability of events which we cannot bring entirely under laws.

Deducing the Probability of a Phenomenon.—There are certain circumstances in which the probability of a phenomenon may be *determined deductively*. For example, we can say at once that in tossing a coin the probability of getting heads is $1/2$; we know that there are two possibilities and only two; if the coin is properly made we know of no reason why one side should, in any particular case or in the long run, fall any oftener than the other.[1]

We say that their chances are equal and that the probability of each is $1/2$. In the case of a die the probability that any specified side will come uppermost is $1/6$. There are six possibilities, all equal. In the long run we expect each side to turn up as frequently as any other, *viz.*, in one-sixth of the cases. The chances of any one are as $1:5$; one for, and five against. If we have a bag containing twenty balls, three of which are white and the rest black, the probability of drawing a white ball is $3/20$. In this instance there are twenty pos-

[1] It is essential in such calculations that there be no known factor favoring a given result more than any other.

sibilities, three of which would give the desired result; drawing a white ball may be brought about by realizing any one of three possibilities; or, out of twenty possibilities, three are favorable. In instances of this sort we have a definitely known number of possibilities, with *no reason to believe that there is anything tending to bring about one rather than another.* The probability of any specified one among them will be expressed by a fraction having 1 as its numerator and the number of possibilities as its denominator. In the case last cited the probability of drawing any particular ball is $1/20$. If among these possibilities any number of them favor the realization of any particular phenomenon, the probability of that phenomenon will be expressed by a fraction having as its denominator the total number of possibilities and as its numerator the number of possibilities favorable to the occurrence of the phenomenon in question. If all the possibilities were favorable (*e. g.*, if all twenty balls were white), the fraction would be $20/20$ or 1, which is the symbol for certainty, or the upper limit of probability; if none were favorable, it would be $0/20$ or 0, the lower limit, or impossibility.

Suppose we toss the coin twice (or toss two coins), what is the probability of getting heads both times? There are four possibilities, as follows:

H H
H T
T H
T T

We might get heads in both, or heads in the first and tails in the second, or tails in the first and heads in the second, or tails in both. In only one of these does heads come in both throws; the probability is therefore $1/4$. It is the same for two tails. For one heads and one tails it is $1/2$. For heads in the first throw and tails in the second it is again $1/4$, and so on.

If we should toss it three times, the probability of getting heads each time would be $1/8$. There are eight possibilities:

H H H	T H H
H H T	T H T
H T T	T T H
H T H	T T T

We can get the probability for any number of throws by multiplying together the probabilities for each of the several throws; for two throws, $1/2 \times 1/2$, or $1/4$; for three, $1/2 \times 1/2 \times 1/2$, or $1/8$; for five, $1/2 \times 1/2 \times 1/2 \times 1/2 \times 1/2$, or $1/32$, and so on. The results would be the same if we should throw several coins at once instead of throwing one several times. Suppose we are drawing balls from two bags; one of them contains three white and seventeen black balls; the other contains two white and eight black balls. What is the probability of drawing a white ball from each? The probability, in one case, is $3/20$, and in the other $2/10$; the probability of getting a white from each is therefore $6/200$ or $3/100$. Each in one bag might be drawn with any one in the other; hence there are 200 possibilities, only six of which are favorable; each of the two whites in one bag might be drawn with each of the three in the other.

The probability of getting any combination of independent events is thus obtained by taking the product of the probabilities of the several events.

If two events are *mutually exclusive* the probability of getting *one or the other* would be *the sum of their independent probabilities.* In tossing a coin, the probability of getting heads is ½ and that of getting tails is the same; the probability of getting one or the other is the sum of the two or 1 = certainty. In throwing a die, the probability of getting a five is ⅙, the probability of getting a six is the same, the probability of getting one or the other is ⅓.

In tossing a coin twice, "It might be argued that since the probability of throwing heads at the first trial is ½ and at the second trial also ½, the probability of throwing it in the first two throws is 1, or certainty. The true result is ¾, or the probability of heads at the first throw, added to the exclusive probability that if it does not come at the first, it will come at the second." [2] The probability that it will come in the first is ½. The probability that it will not come in the first is also ½; the probability that it will come in the second is also ½. The product of the two last gives the probability that if it does not come in the first, it will in the second. This product, added to the first ½, gives the probability that it will come in at least one of the two throws. There are, of course, four possibilities in two throws, and three of them give at least one heads.

If we represent the probability that an event will happen by p, then the probability that it will not hap-

[2] Jevons, *Principles of Science,* chap. x, 3.

pen is $1-p$. The probability in throwing a die that five will not come up is $1 - \frac{1}{6}$ or $\frac{5}{6}$.

Let us suppose a case in which six coins are tossed (or in which one coin is tossed six times); what are the probabilities of 6, 5, 4, 3, 2, 1 and 0 heads respectively? There will be 64 possibilities, as follows:

6	0
H H H H H H	T T T T T T
5	**1**
H H H H H T	H T T T T T
H H H H T H	T H T T T T
H H H T H H	T T H T T T
H H T H H H	T T T H T T
H T H H H H	T T T T H T
T H H H H H	T T T T T H
4	**2**
H H H H T T	H H T T T T
H H H T T H	T H H T T T
H H T T H H	T T H H T T
H T T H H H	T T T H H T
T T H H H H	T T T T H H
H H H T H T	H T H T T T
H H T H T H	T H T H T T
H T H T H H	T T H T H T
T H T H H H	T T T H T H
H H T H H T	H T T H T T
H T H H T H	T H T T H T
T H H T H H	T T H T T H
H T H H H T	H T T T H T
T H H H T H	T H T T T H
T H H H H T	H T T T T H

3	3
H H H T T T	H H T H T T
H H T T T H	H T H T T H
H T T T H H	T H T T H H
T T T H H H	H T H H T T
H H T T H T	T H H T T H
H T T H T H	T H H T H T
T T H T H H	T H T H H T
H T T H H T	T H T H T H
T T H H T H	H T H T H T
T T H H H T	T H H H T T

One combination gives six heads, six give five, etc. The probabilities for 6, 5, 4, 3, 2, 1, and 0 heads are respectively 1/64, 6/64, 15/64, 20/64, 15/64, 6/64, 1/64. In ten throws the number of possibilities would be 1024; the numbers favoring 10, 9, 8, 7, 6, 5, 4, 3, 2, 1, and 0 heads would be respectively, 1, 10, 45, 120, 210, 252, 210, 120, 45, 10, 1. Examination will show that these series of numbers (1, 6, 15, 20, etc., and 1, 10, 45, 120, etc.) are the coefficients of the terms of a binominal raised to the power indicated by the number of throws (6, 10, etc.). In these cases the phenomena are not mutually exclusive. The fact that heads comes (or does not come) in any throw makes no difference to the other throws.

There are important scientific applications of the facts just brought out.[1] "Suppose, for the sake of argument, that all persons were naturally of the equal stature of five feet, but enjoyed during youth seven independent chances of growing one inch in addition.

[1] Jevons, *Principles of Science*, chap. ix, 5.

Of these seven chances, one, two, three, or more, may happen favorably to any individual; . . . out of every 128 people:—

1	person	would	have	the	stature of	5	feet	0	inches.
7	persons	"	"	"	"	5	"	1	"
21	"	"	"	"	"	5	"	2	"
35	"	"	"	"	"	5	"	3	"
35	"	"	"	"	"	5	"	4	"
21	"	"	"	"	"	5	"	5	"
7	"	"	"	"	"	5	"	6	"
1	"	"	"	"	"	5	"	7	"

The probability of a stature of 5 feet 1 inch would be $7/128$, the whole number of possibilities being the denominator.

There are sometimes cases in which *an event has happened in all the instances in which a given circumstance or set of circumstances has been present*—B has always followed A. What is the probability of its happening the next time the circumstance or circumstances recur? Let the number of times it has happened be represented by m. Then the probability of its happening the next time is expressed by the fraction $\frac{m+1}{m+2}$. This may be determined by a series of mathematical operations or more simply as follows: "In this fraction the denominator represents the sum of conceivable cases, since after m real cases have occurred there are always two additional cases, which we can think of as occurring, *viz.*, the repetition or non-repetition of E (the event); the numerator as usual denotes the number of favorable chances. The example usually adduced is that as the alteration of day with night has now been historically attested for 5,000 years, the probability of the same alternation recurring to-day = 1,826,214 : 1,826,215; that is, one may bet 1,826,214 to 1 on its occurring again."[3]

Dangers to be Avoided in Interpreting Probabilities.—Care is needed in the employment and interpretation of probabilities. When we say that the probability of heads is $\frac{1}{2}$, we do not mean that in two throws we shall get heads once, nor do we mean that in any definite number of throws the number of heads and tails will be equal. Indeed, a run of heads of any finite length is possible. The probability of a run of ten is $\frac{1}{1024}$. But the longer the series the more closely should we expect the proportion of heads to approximate that indicated by the statement of its probability. If heads have come up four times already the probability of their coming up the next time is still $\frac{1}{2}$; the same is true if they have come up ten or any other number of times. The past throws have nothing to do with the present or future. To expect that, because a coin has come up heads several times in succession, it is therefore more likely to come up tails the next time,[4] is wholly to misunderstand the meaning of probability. Indeed, a preponderance of heads in the past throws would suggest that the coin was not true, that there was a hidden cause favoring heads, and that as a matter of fact the probability of heads was greater than $\frac{1}{2}$. In no case does the knowledge of the probability of an event give any definite information regarding the next or any other

[3] Lotze, *Logic,* Book II, chap. 9.
[4] This is sometimes called the "gambler's fallacy."

specified case. It simply tells us that in the long run we have a right to expect a certain proportion of occurrences, and the longer the run, the closer the approximation. So far as "luck," pure and simple (favorable or unfavorable accident), is concerned, we might expect that, in the long run, and taking an infinite number of individuals, "good luck" and "bad luck" would be equal, though particular individuals might be always lucky or unlucky, or their good or bad luck might begin or end at any point whatever. Bad luck in the past would be no evidence of bad luck in the future. Of course, a great deal that we ordinarily call luck or chance is really the result of ability, foresight, and so on, or of their opposites.

EXERCISES.

1. What does each of the following propositions mean?
   (1) He is probably an official of some sort.
   (2) There is a strong probability in favor of his being elected.
   (3) The probability of ten more years of life for a man of his age is $\frac{1}{6}$.

2. "The probability that a new-born child will live to the age of 25 years is $\frac{1}{5}$; and if it lives to that age, the probability of its being well-educated is $\frac{1}{4}$; and if it is well-educated, the probability of its being a distinguished person is $\frac{1}{20}$. Calculate the probability of the new-born child's being a distinguished person."—Ray, *Logic.*

3. Thirty per cent. of the men in college are Freshmen, and twenty per cent. of the Freshmen come from the West. What is the probability that the next student who passes will be a Freshman from the West? A Freshman and not from the West? Not a Freshman?

4. What is the probability that a die will fall with the same side up five times in succession?

5. What is the probability that a given side of a die will come up at least once in two throws? In three throws?

6. In tossing a coin four times, what is the probability of getting heads three times?

7. Suppose that an event had happened in one thousand cases in which a given phenomenon had been present and had never failed to happen when the latter was present; what would be the probability of its happening when next the phenomenon was present?

SUPPLEMENT TO PART II

THE GRAPHIC METHOD OF REPRESENTING DATA AND THEIR RELATIONS

When large groups of figures are to be presented it is often useful to employ diagrams which will enable the eye to grasp at once the series as a whole. There are many varieties. Popular discussions of comparative populations, wealth, navies, and so on, often represent the various figures by lines or surfaces which are so juxtaposed as to show at once to the eye the relations of the several quantities. Thus, the tonnage of the eight greatest navies of the world in 1907 was approximately

G. Britain . . 1,633,000
U. S. 611,500
France 609,000
Germany 529,000
Japan 374,000
Russia 207,500
Austria 113,000

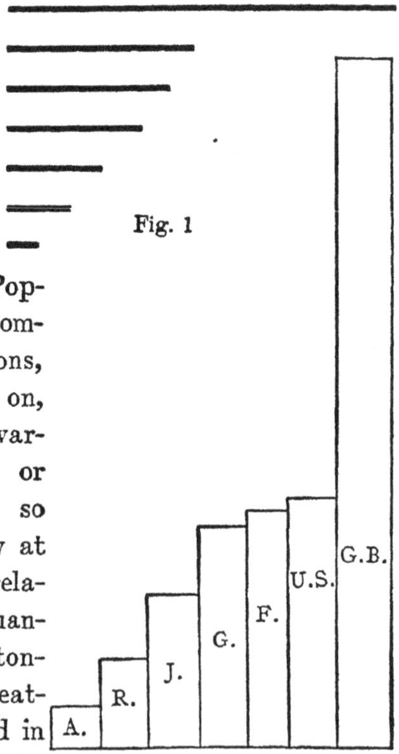

Fig. 1

Fig. 2

as above. Or we might employ rectangles with equal bases, or points on a curve.

The relation between two series of numbers or quantities may be represented graphically. Let us take a case in which successive quantities are related to successive years. The population of the United States at each census, from 1820 to 1900, was approximately: 1820, 9 millions; 1830, 12 millions; 1840, 17 millions; 1850, 23 millions; 1860, 31 millions; 1870, 38 millions; 1880, 50 millions; 1890, 62 millions; 1900, 76 millions. These facts could be represented thus:

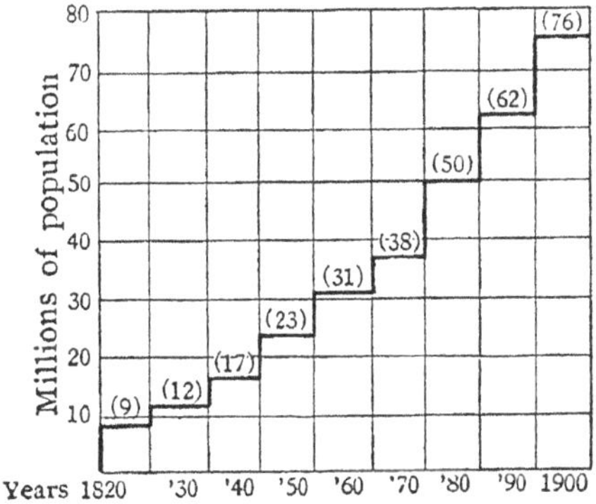

FIG. 3.

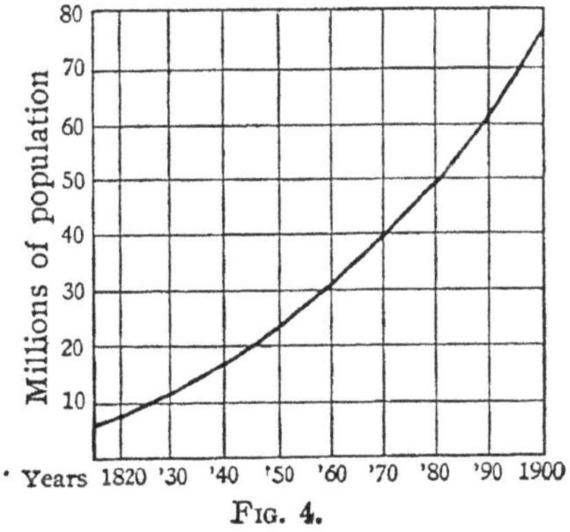

FIG. 4.

In the first figure the successive quantities are represented by rectangles with equal bases, the years being indicated on the base line and the population on the

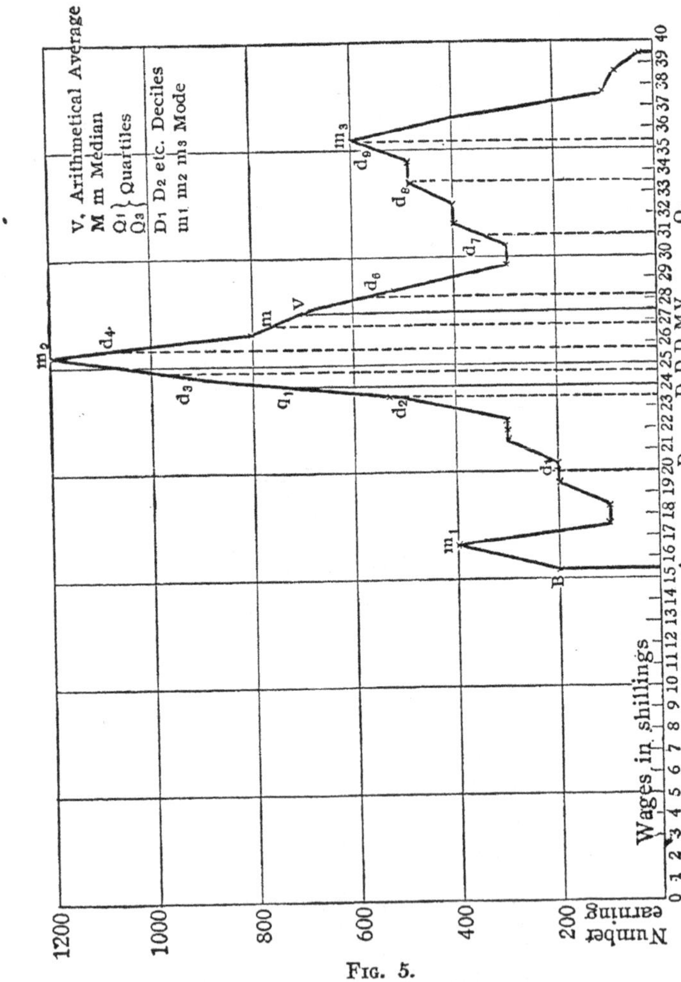

FIG. 5.

vertical. Or we can indicate the quantities by a series of points located according to their size and date and join these points by a curve, as in figure 4.

Figure 5 is borrowed from Bowley's *Elements of Statistics*, and represents the following data:

	SHILLINGS	
Numbers of work people earning from	15 to 16 —	200
"	16 to 17 —	400
"	17 to 18 —	100
"	18 to 19 —	100
"	19 to 20 —	200
"	20 to 21 —	200
"	21 to 22 —	300
"	22 to 23 —	300
"	23 to 24 —	500
"	24 to 25 —	900
"	25 to 26 —	1200
"	26 to 27 —	800
"	27 to 28 —	700
"	28 to 29 —	500
"	29 to 30 —	300
"	30 to 31 —	300
"	31 to 32 —	400
"	32 to 33 —	400
"	33 to 34 —	500
"	34 to 35 —	500
"	35 to 36 —	600
"	36 to 37 —	400
"	37 to 38 —	100
"	38 to 39 —	80
"	39 to 40 —	20

Median, 26/9 (26 shillings, 9d); Quartiles, 24/2, 32. Deciles, 20, 23/6, 24/9, 25/8, 26/9, 28/2, 31, 33/4, 35/4.

Mode, 25/3; secondary positions (modes), 16/6, 36.

Curves may also be employed to show the relations between two or more sets of variable quantities. Thus:

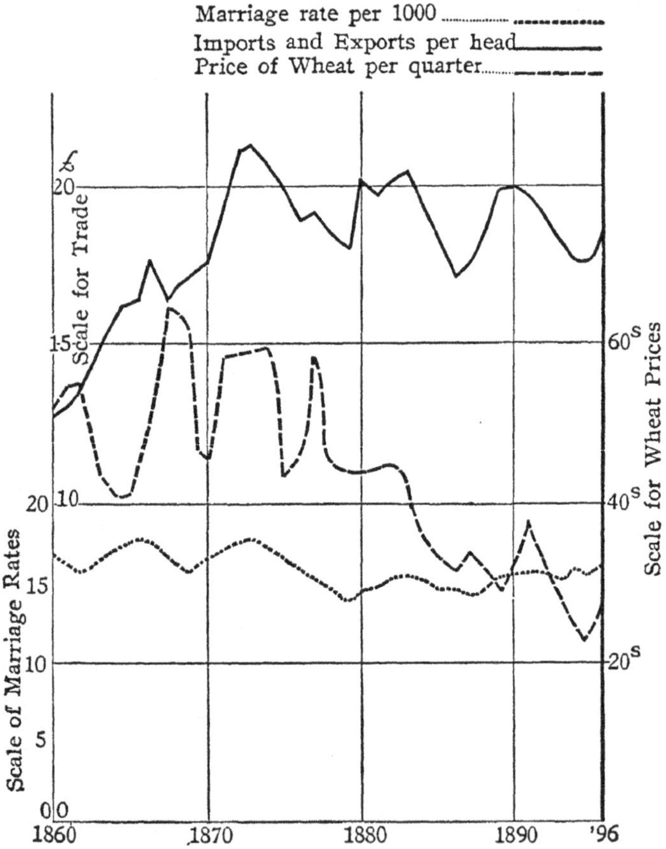

FIG. 6.

There are many opportunities for error in such a case as this. For example if the scale for the marriage rate were twice as great, the fluctuations in the curve would appear to be greater in comparison with those in the other curves. (See Bowley, *loc. cit.*) Great caution is necessary in the comparison of curves.

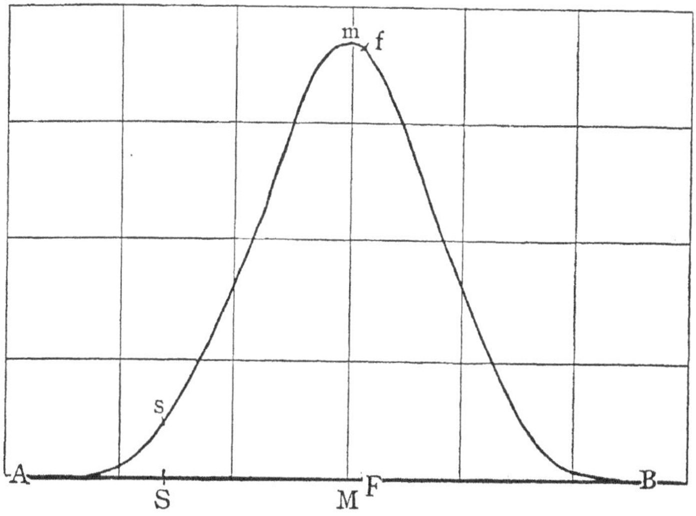

Fig. 7.

The Probability Curve.—The curve represented by Fig. 7 is of great value in scientific investigations. It is sometimes known as the Probability Curve, or the Normal Curve of Error. Suppose a large number of measurements of a given quantity, or variations from a given standard; if no causes of constant error are present, errors of excess are equally probable with errors of the opposite sort as we have already seen (page 210); moreover large errors are less probable than small errors and very large errors are very improbable. Let the line *Mm* represent the correct measurement; let the part of the curve to the right of this line represent the positive errors (errors in excess), and the part to the left, those which are negative; the size of an error would be indicated by its distance from *M* along the base line. *MF* would indicate a small positive error; *MS*, a large negative one; height above the base line indicates the comparative number of the errors; thus the line *Ff* means a large number of small errors; the line *Ss*, a small number of large ones. The curve may represent not only errors, but any series of quantities grouped about a type, when the causes of variation are very numerous and are independent of each other. In an earlier example it appeared that in tossing a coin the various possible series of runs of heads or tails could be stated in a series of figures which were related to each other as are the coefficients in the expansion of a binomial to the power indicated by the number of throws.

This formula can be used in any case in which a number of independent variable factors is concerned. If the number is very large the chances of the various possible combinations can be represented by the curve we are discussing, as for example, in the case of the stature of adult males. Many factors enter into the determination of stature, such as heredity, health in childhood, kind and quality of food, occupation, and so on; the statures of men in any given community are ranged on either side of the mode in such a way as to be represented with substantial accuracy by the probability curve (or the Curve of Frequency or of Distribution, or the Normal Curve of Error, as it is variously called).

Let us suppose, however, that some constant factor is introduced tending to alter stature in a given direction; what effect will this have on the curve? It will obviously change its form, for a greater number of cases will appear on one side of the former mode and a smaller number on the other; instead of being symmetrical it will be skewed to one side or the other, thus:

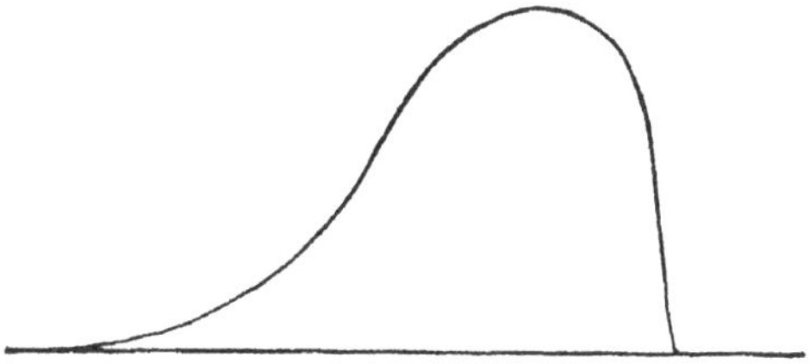

FIG. 8.

The presence of a skew always indicates the operation of some constant factor. If in tossing coins we found such a skew toward the side representing a large number of heads we should have evidence of the presence of some constant factor favoring heads.

Sometimes the representation of a series of quantities would produce a curve of still more irregular shape, such as this:

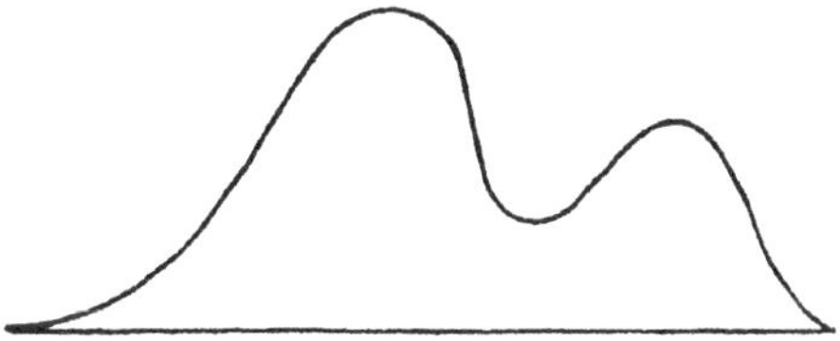

FIG. 9.

Such a curve would show that the group was not homogeneous; that there was really a combination of two groups; or that certain factors were operative in one part of the series and not in another. Professor

Thorndike[1] gives the following curve as representative of the frequency of death at different ages, the age

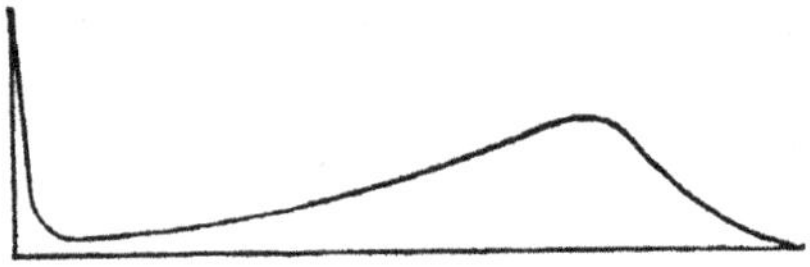

FIG. 10.

increasing as we go from left to right. Certain factors are operative in early infancy that play no part later.

1 In his *Theory of Mental and Social Measurements*, p. 51. See this book for discussion of measurement.

PART III

THE CONSTRUCTION OF SYSTEMS

CHAPTER I

EXPLANATION

What is Explanation?–Probability, classification and the discovery of laws all have to do with facts. Probability tells us the frequency with which a fact may be expected to occur; classification puts the fact into a group of like facts, and the better the classification from a scientific point of view, the more does the placing of the particular fact tell us with regard to its relations of resemblance and difference with other facts. A law states the conditions under which a fact occurs. In which of these cases can we be said to *explain* a fact?

The statement of the frequency with which an event occurs does not explain the event. We may say that the order of nature is such that, unless some change in the conditions is introduced, we may expect an event to occur with the same frequency in the future as it has manifested in the past; but that is obviously very far from an adequate explanation.

Do we explain an event when we classify it? When we ask why a given body fell to the ground, do we explain the phenomenon by saying that it was a heavy body? Not entirely, and if we did not already know some law holding for heavy bodies, our statement would throw no light on the subject. It is quite true that a statement of this sort may be a preliminary to explanation.

A law tells us how phenomena *of a given sort* behave; it states the conditions of their occurrence, and if we can not say what sort of thing a given fact is we can not state its conditions. In bringing a fact under a law we first approach an explanation of it. Explanation has been defined as (in positive science) "*the reduction of a phenomenon to the terms of a general principle, whatever that principle may be.*"[1]

Have we reached a final and complete explanation of a fact when we have brought it under a general principle? In many cases this seems to be sufficient; if we are familiar with the law and if we can see its bearing upon the fact in question, we are ordinarily content with this sort of explanation. An eclipse of the moon is sufficiently explained for ordinary purposes if we are told that it is caused by the presence of an opaque object between it and the source of its light. Or the revolution of the moon about the earth may be explained by saying that it is the resultant of the operation of centripetal and centrifugal forces.

But there are two further questions that may be asked. First, *what are the circumstances in which the law operates in the present case?* and second, *how is the law itself to be accounted for?* Let us consider the second question: How is a law to be explained? The answer is: By showing that the law is itself a case of a more general law. The attraction of the earth for bodies on its surface is explained by showing that it is a case under the law of gravitation. "It has often been found that scientific men were in possession of

[1] *Dictionary of Philosophy,* etc. Ed. Professor J. Mark Baldwin.

several well-known laws without perceiving the bond which connected them together. Men, for instance, had long known that all heavy bodies tended to fall towards the earth, and before the time of Newton it was known to Hooke, Huyghens and others, that some force probably connected the earth with the sun and moon. It was Newton, however, who clearly brought these and many other facts under one general law, so that each fact or less general law throws light upon every other."[2]

How far can this be carried? Do we not at last arrive at laws which are elementary and not to be explained by reference to anything simpler or more fundamental? Are we then to regard these elementary laws as inexplicable? No, for *reference to simpler and more fundamental laws is merely one method of bringing the data into a system.* If the elementary law can be shown to be a part of a system, made up, for example, of other elementary laws, we have all the explanation which can be demanded. The parts are explained by being given their proper place in the whole. If the whole were in turn a part of a larger whole it could be explained in the same way. But suppose the whole were ultimate: could it be explained? Could the universe,—not simply the physical universe, but the whole of reality of whatever kind,—could it be explained? The only kind of explanation which could be given would be a statement of the relation between this whole and its parts, and it is hard to see what other kind could be asked for. One might ask for the purpose of it all, but in the broad way in which we are conceiving

[2] Jevons, *Lessons in Logic,* p. 268.

of it, whatever purpose existed would be included in the whole itself.[3]

The explanation of a law (or fact) involves not merely showing that it is a case under a general rule, but also in discovering its relations to other laws and facts aside from the general rule under which it is brought. Along with other laws and facts it constitutes an interrelated whole, and explanation might better be defined as giving the thing to be explained its place in an organized system. We may wish to know not only the general law or laws under which the special case falls, but also the other special cases which are related to it; or in other words, we may wish to know the general law and also the circumstances of its application in the present case.

We can distinguish between *explanation in general terms* and *specific explanation.* The first consists in assigning the general law; the second in showing, in addition to this, how the general law applies. The explanation of the moon's eclipse as given above is of the first sort; if we should supplement it by stating that the intervening body is the earth and show how the earth and moon are situated with reference to the sun, we should have a nearer approach to a specific explanation. If we should go further and give an account of the way in which they happened to be in these positions at this particular time we should have a still more complete and specific explanation. The explanation of a fact, to be absolutely complete and specific, would have to contain a statement of all the laws of

[3] See Hobhouse, *The Theory of Knowledge,* chapter on "Explanation."

the system to which the fact belonged and also a description of all the facts contained in the system. Or otherwise expressed, a complete specific explanation would present all the facts which were related to the one to be explained, together with the laws under which each and every one of these relations fell, in a single unified whole. If an explanation is complete we can start with the laws and circumstances involved and show that the phenomenon to be explained would necessarily follow from them. Such an explanation might be possible in some parts of the field of astronomy; if not an absolutely complete explanation, it could be sufficiently complete for all ordinary purposes. Complete grasp of a system would enable us to reconstruct the past or predict the future of the system. The astronomer's knowledge of the solar system is so thorough that, starting from the present position of any member of the solar system and knowing the relation in which it stands to each of the others, he is able to calculate the present position of any of the others and also to reconstruct their relative positions in the past or to predict their positions in the future; he can state the number and dates of eclipses which took place a thousand years ago and he can calculate the relations of any member of the system to any other at any given time. The only limits to such calculations are those which result from an imperfect knowledge of the solar system itself and ignorance of many possible disturbances from without. If a system were entirely isolated and if its laws were completely known, then a description of the system at any given time would make possible a description of it at any other time. Needless to say there is no system of facts in nature which is entirely independent of the rest. Every system of facts is related to every other system, either directly or indirectly. To get a complete explanation of any single fact it would be necessary to know the whole system of nature. To a knowledge of all the laws of nature and all their relations to each other, it would be necessary to add a complete description of the relations of the parts of the system to each other. Then we could have an explanation of the fact which would show its relations to the whole and to every part of the whole. Obviously, an explanation of this sort can not be given for any fact. If it could be given for one it could be given for all. In this sense, Tennyson's lines are literally true:

> "Flower in the crannied wall,
> . . . . .
> If I could understand
> What you are, root and all, and all in all,
> I should know what God and man is."

Sometimes we are satisfied with an explanation in general terms, in terms of some familiar law. Sometimes we call for an explanation of the law and are satisfied if it can be referred to some more general law or to some familiar system.[4]

[4] Each of the "explanatory" sciences deals with some limited group of facts and attempts to discover the system of laws which holds within that field. Every one of the positive sciences takes for granted certain general principles, valid for all knowledge, and tries to discover a system of laws which shall be valid for its own field. A demand for the explanation of a physical fact, which is not satisfied with a statement of its relation to other physical facts in terms of physical laws, is not a problem for the science of physics. Ultimate explanations are for an ultimate science; philosophy is sometimes defined as this ultimate science. The laws which underly all the other sciences would be the province

The aim of most of the sciences is not to explain particular facts but to furnish the general laws, which may be employed in explanation of facts in a given field. A science like mechanics, for example, "lays down propositions which are true in the same way of all fluids, all gases, etc., and represents them as general consequences of general presuppositions." "It does not descend into the whole manifold of the given, inasmuch as it deals only with events which take place in a similar manner in bodies differing in many other respects. That this or that phenomenon falls under these laws is a matter for subsumption in dealing with the particular; it is not necessary to the completeness of the arrangement as a whole. The mechanical theory of gases disregards their chemical differences in so far as they do not affect its special province by giving rise to differences of specific gravity; it is no part of its task to enumerate how many sorts of gases there are: it is enough to say that if a body is a gas it conforms to certain laws of compressibility, of expansion by heat, or of capacity for heat, etc." [5]

Corresponding statements might be made about any one of the sciences which aim at the discovery of laws (as distinguished from the classificatory sciences, which aim only at the complete classification of facts).

of such a universal science. It would be related to the particular sciences as they are related to the concrete data which they investigate. It would not be complete till each of them was complete, nor could any of them be final until it had reached its goal; just as, in any one of them, no fact is completely known unless the general body of laws is known, and the body of laws can not be completely known until every fact can be accounted for. Progress consists in alternate advancements toward completeness along both lines, from both directions.

[5] See Sigwart, *Logic,* Part III, chap. vi.

Specific Explanation.—But in practical life and in the historical sciences we are concerned with the concrete individual fact; hence propositions which deal with general properties and the like are of no use to us unless we can see how they apply in the particular case; and usually we cannot see that unless we have independent knowledge of the attendant circumstances. The knowledge of the single fact and of the general law under which it falls does not ordinarily enable us to reconstruct the system; our knowledge of both fact and laws is too incomplete to enable us to get a specific explanation; and a general explanation is not sufficient as a guide for action.

Suppose that the fact under investigation were a criminal act of some sort; the object of the prosecuting attorney is to fix the responsibility for the act. In order to do this it is necessary for him to reconstruct the circumstances surrounding the doing of the act. In other words, he must build up the system of concrete facts to which the act belongs. A Sherlock Holmes might be able, by superior powers of observation, unusual knowledge of the laws of criminal behavior, and so on, and by extraordinary skill in bringing each fact observed under the proper law, to reconstruct the whole system of facts without reliance upon the testimony of others; but ordinary mortals would usually find it necessary to collect evidence from all possible sources, and then, out of the scattered bits, to restore some semblance of the original whole. One fact, sufficiently described, and a thorough grasp of the principles of the system, might be enough; lacking that, as many fragments as possible must be collected, and even then no certain result might be reached.

Knowledge of the laws alone does not ordinarily suffice, and even if that knowledge were complete it might be possible to go astray in applying it. The greatest scientist might fail in the attempt to give a specific explanation of a concrete fact if he should overlook the circumstances surrounding the fact. Specific explanation is a matter for the *application* of science, and theory alone is proverbially insufficient as a guide for practice, no matter how correct the theory may be in the abstract. The "theorist" may have an incorrect theory or he may fail to note the special circumstances in applying one which is correct.

Closely related to the processes of explanation are those of prediction. They are complementary. In explaining, we give the laws and circumstances which account for a fact or a supposed fact; in prediction we set out from a set of laws and circumstances and attempt to show that a certain fact will occur in the future or will be found on further investigation to exist. Successful prediction is, as we have seen in studying the inductive methods, a test of the validity of the laws we are employing. If we do not know the laws and circumstances we can not predict successfully, except occasionally and by accident. Prediction may be said to represent the practical application of science.

CHAPTER II

HYPOTHESIS

What is an Hypothesis?—Before we go on to the discussion of certain typical systems, it is necessary to give some attention to another matter involved in most explanations, namely, the use of hypotheses. The term hypothesis is used in several different senses, but for our purposes an hypothesis is a *provisional explanation*. Fictions made for the purpose of argument, illustration, or simplification may also be regarded as hypotheses, but we shall use the term in the narrower sense.[1]

Hypotheses, or provisional explanations, may assert the existence of a fact, as when we assume that a defective flue was the cause of a fire; or of a law, as when we infer a causal connection between vaccination and freedom from smallpox; or of a complex system of laws and facts, as when we infer the existence of a matriarchal system in the early history of certain peoples. Inductive inferences, which were discussed in chapter vi., would fall under the head of hypotheses.

The Value of Hypotheses.—There has been much disagreement regarding the value of hypotheses and their use in science. A good many scientists have declared that hypotheses are not only unnecessary but

[1] For an interesting discussion of the subject, see Muirhead, *Philosophy and Life,* Art. "Hypothesis."

are positively harmful, and Newton's "*Hypotheses non fingo*" is often quoted in defense of their position. To apply this literally would mean that a science would remain merely a body of carefully observed and classified facts, unless laws should somehow or other spring ready made from them without having been previously put forward as possible laws and then tested by further observation and experiment. Now, it might sometimes be possible to collect our facts over a very wide range and classify them and their relations in such a way as to show at once the law of their connections; the inductive methods indicate the sort of grouping that would be necessary; but even with their assistance it would rarely happen that a fully verified law would appear without previous unsuccessful attempts at its discovery. Previous to the Nineteenth Century the progress of science was seriously retarded by what Romanes has called the Bugbear of Speculation. In the introduction to his *Darwin and after Darwin* he gives the following statement of the situation in the natural sciences: "Fully awakened to the dangers of web-spinning from the ever fertile resources of their own inner consciousness, naturalists became more and more convinced that their science ought to consist in a mere observation of facts, or tabulation of phenomena without attempt at theorization upon their philosophical import. If the facts and phenomena presented any such import, that was an affair of the man of letters to deal with; but as men of science, it was their duty to avoid the seductive temptations of the world, the flesh and the devil, in the form of speculation, deduction, and generalization . . . this was the orthodox and general view." It was current in the time of Linnæus and even to the time of Cuvier. "*The Origin of Species* made an epoch . . . Darwin displayed the true principle which ought to govern biological research . . . he never loses sight of the distinction between fact and theory . . . but his idea of the scientific use of facts is plainly that of furnishing legitimate material for the construction of theories . . . the spirit of speculation is the same as the spirit of science, namely, to know the causes of things . . . If it be causes or principles as distinguished from facts or phenomena, that constitute the final aim of scientific research, obviously the advancement of such research can be attained only by the framing of hypotheses. And to frame such hypotheses is to speculate." Darwin said of himself that he made an hypothesis on every subject. "He was as productive of hypotheses as nature is of living things, and like her, he subjected them all to the principle of natural selection." [2]

Hypotheses are necessary for science. "All science starts with hypotheses, in other words, with assumptions that are unproved, while they may be and often are erroneous; but which are better than nothing to the seeker after order in the maze of phenomena." [3]

An erroneous hypothesis may be quite as effective in the field of practical activity as a true one could be. "The theory that some god would destroy the tribe if it did not wash at a particular time was a very crude explanation of an observed fact; but it nevertheless has its merits. It caused the tribe to wash occasionally—

2 Cramer, *The Method of Darwin*, p. 40.
3 Huxley, *Hume*, p. 65.

a thing it would never have done. It furnished a theory which tended to prevent disease. It recognized the truth which bacteriological science has only just grown up to in the present generation: that the penalty for violation of law was visited not so much on the individual as on the community.[4]

The Ptolemaic System was an erroneous hypothesis, but without it, or some other theory, astronomical knowledge would have progressed much more slowly than it did. "The superiority of the Greeks to their Oriental neighbors in science has often been accounted for by their fertility in theory. The Oriental peoples were, at the time of which we write (Cosmological Period), considerably richer than the Greeks in accumulated facts, though these facts had certainly not been observed for any scientific purpose, and their possession never suggested a revision of the primitive view of the world."[5]

The danger in using hypotheses lies in the fact that we are so likely to forget that they are only hypotheses. We find some explanation which seems to fit the facts or which supports some other belief of ours, and we forget that our hypothesis has not been verified. We tend to have too great a fondness for hypotheses which we have ourselves made; we are liable to "the partiality of intellectual parentage." Darwin's example may again be cited as the right one. "I have steadily endeavored to keep my mind free so as to give up any hypothesis, however much beloved (and I cannot resist

4 President Hadley, in an article in the *Atlantic Monthly*, February, 1903, p. 153.

5 Burnet, *Early Greek Philosophy*, 1st Ed., p. 22.

forming one on every subject), as soon as the facts are shown to be opposed to it. Indeed, I have had no choice but to act in this manner, for, with the exception of the Coral Reefs, I cannot remember a single first-formed hypothesis which had not after a short time to be given up or greatly modified."[6]

It has been said that an hypothesis is a question.[7] When we form an hypothesis our attitude should be described by the inquiry: "Is this the true explanation of the facts?" If that attitude could be preserved the danger from hypotheses would be very small. We often hear of "working hypotheses." They are simply hypotheses which are confessedly unverified but valuable as a basis on which to work toward an explanation. There may be many degrees of verification, from the most complete to the most imperfect; strictly speaking, one might say that until an hypothesis has been proved true and shown to be a law (and therefore no longer a mere hypothesis) it remains a working hypothesis; but in ordinary usage this term is used to describe hypotheses which are useful but in no sense established.

One way of holding the mind open to the fact that a given hypothesis is not to be trusted too far would be to keep before us a number of different rival hypotheses.[8]

[6] *Life and Letters,* Vol. I, p. 83, quoted in Cramer's *Method of Darwin.*

[7] Langlois and Seignobos, *Introduction to the Study of History.*

[8] This method, which has been called the "Method of Multiple Working Hypotheses," has been recommended as promoting thoroughness, suggesting lines of inquiry, as a means of sharpening discrimination, increasing fertility in reasoning processes, etc. See Chamberlain, *The Method of Multiple Working Hypotheses, Science,* Feb. 7, 1890.

How are Hypotheses Suggested to Us?—We have already seen that the groupings of facts or sequences such as we use in applying the method of Agreement and the rest lead to the formation of inductive hypotheses. Indeed (1) *any sequence* may do this. If we notice that A is followed by B, we tend naturally, in the absence of evidence to the contrary, to believe that the second will always follow the first. The statement that these two are universally and necessarily connected in this way is an hypothesis unless or until further examination shows that the statement is either true or false.

(2) ANALOGY.—A second[9] source of hypothesis is found in *Analogy*. The term analogy has been used in a good many different senses.[10] In its broadest sense it means any kind of resemblance. An inference from analogy is inference from the resemblance of two cases in certain observed points to their resemblance in a further particular which has been observed in only one of them. For example, we may observe, in examining the skulls of certain extinct animals, that they all have sharp canine teeth and rudimentary molars; in this respect they resemble modern carnivorous animals; hence we infer that these extinct animals were carnivorous. Or we know that a boy who is ill has eaten unripe fruit; we infer that another boy who shows similar symptoms has been guilty of a similar indiscretion. Inference from analogy is usually to some particular

9 A third source of hypotheses is found by Sigwart in the Conversion of propositions. For example, isosceles triangles have the angles at the base equal. Is the converse true? See Sigwart, *Logic,* ii, p. 83.

10 See Minto, *Logic,* p. 367.

fact or situation, though it may be further extended to a general principle.

Analogy alone is notoriously an unsafe guide; but if certain *general rules* are kept in mind it may often be employed to advantage.

1. If two sets of facts resembled each other in only one particular or in very few, an inference as to their resemblance in another particular would be very hazardous. The fact that two men were born in the same city gives no ground for the conclusion that one will have the same profession as the other.

2. If two sets of facts resembled each other in every particular except such as were irrelevant, the inference would be safe. Reasoning by analogy in geometry illustrates this. Again, if two animals were alike except in color, the fact that one was carnivorous would be good ground for believing that the other was too.

3. In so far as two things or two sets of facts differed in relevant particulars the inference would be of doubtful value. If one of two twins had been educated in one way and the other in a different way, it would not be safe to infer that one would be interested in the things in which the other was interested.

4. If one of two similar things possessed a characteristic inconsistent with a characteristic possessed by the other, it would of course be impossible to infer that the first thing possessed the latter characteristic. If A is a singer we cannot infer that his twin brother B is, if the latter is deaf and dumb.

5. If the points of resemblance outnumbered the points of difference, we should have more reasons for than against the inference, provided, of course, that

the various points were equally important in determining the character of the things in question. As a matter of fact they never are of equal importance, so that the relative importance of the various characteristics should be taken into account; a difference in health would count more against equal strength in two men than similarity of stature, weight and age would count in its favor.

6. In counting resemblances and differences, only those which are independent should be counted; the fact that a planet possesses atmosphere and that it refracts light passing near its surface are not independent; to count these as two points of likeness when trying to find ground for the conclusion that one planet resembled another in any particular respect would be incorrect.

In practice it is not easy to say just what characteristics are relevant or to be sure whether two points are independent.[11]

Analogy may be employed in connection with other grounds of inference; for example, if a given situation possesses factors which are partly like and partly unlike those in another situation we might sometimes infer the presence in the second of some further characteristic possessed by the first, but in less degree. One man might exhibit some of the symptoms exhibited by another who was known to have taken a certain drug; we might infer that the first had taken a smaller quantity of the same drug, and so on.

Requisites of a Good Hypothesis.—Having made our hypothesis on whatever ground, we should ask our-

[11] See Hobhouse, *Theory of Knowledge,* page 289, *seq.*

selves whether it is worthy of serious consideration. To put the question in its usual form, What are the requisites of a good hypothesis?

1. In the first place it must serve the purpose for which it is made; it must offer an explanation for data which have not previously been explained correctly. Sometimes certain other explanations may have been offered and found insufficient; sometimes the data may have been entirely unexplained. An hypothesis which does not connect at least two facts hitherto not properly connected is worthless.

2. A good hypothesis must, of course, be consistent with itself and with all the data concerned. It is sometimes said that an hypothesis must not contradict known laws; if there are laws which are completely verified, the hypothesis must be in agreement with them.

That does not mean that an hypothesis must not disagree with any principles which have been hitherto accepted. Such reasoning would have ruled out the Copernican Hypothesis, and it did, as a matter of fact, lead many to reject that theory. Strictly speaking, any hypothesis which offers a consistent account of the data and their relations has a claim to consideration. In practical life we are often warranted in neglecting new theories which contravene accepted principles, at least until they have been shown to approach in value and soundness those already current; but the specialist in any field is not justified in rejecting a theory simply on the ground that it disagrees with those he has held in the past. It is his duty to test all of them.

3. An hypothesis, to be worthy of consideration, must be capable of verification. If data for its verifi-

cation are not already at hand they must at least be conceivable and their discovery must be within the bounds of possibility. Herodotus, in discussing the various theories of the rise of the Nile, says of the one which connected it with the mythical stream of Ocean: "The person who speaks about the Ocean, since he has transported the question to the dominion of the inscrutable, does not admit of refutation." [12]

4. Other things being equal, choose the simplest hypothesis.

Making hypotheses involves mental activities which go beyond perception and memory. They are often discussed under the heading of "Imagination." But imagination in this sense is not simply imagination in the limited sense of making mental pictures. It involves *constructive* activities often of a highly complicated sort. As "creative" imagination it differs from that of the poet in that it does not have to do necessarily nor primarily with concrete experiences.[13]

EXERCISES.

State the ground of the hypothesis in each of the following examples and estimate the value of the hypothesis:

1. Geologists, watching at what rate changes are occurring in the earth's surface at the present time,—*e. g.*, making of valleys, glacier movements, etc.,—determine the length of time it must have taken to produce the corresponding changes during the so-called geological periods.

2. In looking at the pictures in an art gallery, our attention is specially attracted by one picture whose characteristics impress themselves in our mind. Years afterward, in another country, we again see those characteristics in another picture and we feel certain that both pictures are the work of the same artist.

[12] Gomperz, *Greek Thinkers,* Bk. III. chap. p. 6.
[13] Minto, *Logic,* pp. 335, 336.

3. Cutting tools have edges and places for handles. These flints have edges and places for handles; they are therefore, cutting tools.

4. Some Northwest Coast Indians after seeing and hearing a phonograph for the first time, were asked what they thought it was. Their answer was that it was a very powerful echo which the white man controlled by means of a "strong medicine" or magic.

5. The theory that many philologists hold, that many of the languages of the world may be traced back to a common stock, known as the Aryan, is based on analogy. In Persian, Greek, Sanscrit, etc., several very simple words, usually verbs, such as to give and to be, are found to have almost identically the same root, from which resemblances the common descent is argued.

6. Certain mountains, which have large deposits of basalt, contain gold. When large deposits of basalt are found in other mountains, we may suppose that they also contain gold. If gold is not found, tin is. There seems to be a relation between deposits of basalt and deposits of gold.

7. Noting that certain substances expand when they crystallize, and noting also that certain other substances expand when heated, I might infer that heat causes the latter substance to crystallize and hence to expand.

8. Since ether has been offered as the medium of transmission of light-waves, and since some forms of electricity are forms of wave-motion, we might say that ether is the medium of transmission of electricity.

9. The U. S. is a republic and its citizens are prosperous and contented; we may therefore infer that if Cuba were a republic, her citizens would be prosperous and happy too.

10. Hydrochloric acid turns blue litmus paper red; sulphuric acid has similar properties, and we may infer that it, too, will turn blue litmus paper red.

11. Bones resembling those of an elephant were found in a given locality. We conclude that, at some time or other, elephants lived in this locality.

12. Cotton is grown in the U. S. in a moist, warm climate and a sandy soil; we may infer that Egypt, which has these characteristics, will also grow cotton.

CHAPTER III

TYPICAL SYSTEMS OF KNOWLEDGE

An examination of the methods employed in establishing certain typical varieties of systems of knowledge may help to make clearer the complexity of knowledge and the relations of the processes involved in getting it. Every system, as we have seen, contains laws. Some of them are systems of laws and general concepts; others include also concrete facts. Let us take as an example of the first, the sort of system which is to be found in mathematics or mechanics; and as examples of the second, the system of related facts which the historian or the criminal lawyer aims to establish. The other sciences lie between.

We cannot, in our present discussion, begin at the very beginning. We must grant to the historian and the lawyer the generally accepted laws of human behavior, the accepted principles of science, in short, the working materials of his science. To the mathematician, we must grant his concepts and axioms and postulates, and to both the general principles of scientific method. We wish merely to see how each employs these principles, what his method is. All these concepts and principles have been brought to light in the course of human experience. The mathematician employs chiefly the processes of analysis and deductive reasoning. Observation, testimony and, in general, the means for knowing the *concrete* are, for the most part, left aside in his work.

The Geometric System.—Let us take an example of scientific method as it appears in the science of geometry. Other fields of mathematics differ from this in important respects, but for the purposes of illustration geometry will be sufficiently representative. What is the starting point in geometry and what sort of system does it attempt to build?[1] Geometry starts, not with perceived objects as the natural sciences do, but with a set of concepts and propositions. Among its concepts are those of point, line, magnitude, equality, and so on. Some of these are definable in terms of the others, as "a point is that which has no magnitude." There remain, however, certain concepts which are indefinable, *viz.*, those by means of which all the others are defined. Of these concepts there are two kinds: concepts of elements, and concepts of relations. Besides these concepts geometry has among its data certain propositions which express the relations which hold among its elements. These propositions are known as axioms and postulates.[2]

1 See Oswald Veblen, *Popular Science Monthly*, Vol. LXVIII, Art. "The Foundations of Geometry"; and *Transactions of the American Mathematical Society*, Vol. 5, No. 3, Art. "A System of Axioms for Geometry."

2 No clear line of distinction was drawn by Euclid between axioms and postulates. Both were regarded as unproved and unprovable propositions which must be admitted as true by every one who understood them, as *a priori* truths. At the present day their *a priori* character is very widely questioned, but they are unprovable in that they can not be deduced from any simpler propositions. One way of distinguishing them was to define the axioms as *common* notions and the postulates as *geometrical* premises which must be taken for granted. But the line was not clearly drawn and propositions which sometimes appeared as postulates were at other times put among the axioms.

As an axiom we may cite the first in the list: "Things which are equal to the same thing are equal to one another"; and as a postulate: "A straight line may be drawn from any one point to any other point." "All right angles are equal to one another" has sometimes been classed as an axiom and sometimes as a postulate. Euclidian geometry may be defined as "a system of propositions codifying in a definite way our spatial judgments." Every one of its propositions can be deduced from its axioms and postulates, excepting, of course, the axioms and postulates themselves. To prove any proposition we have simply to combine certain concepts and propositions into a coherent whole.

Let us examine Proposition XV, Book I. "Where two straight lines (AB, CD) intersect each other, the vertically opposite angles made by them, are equal."

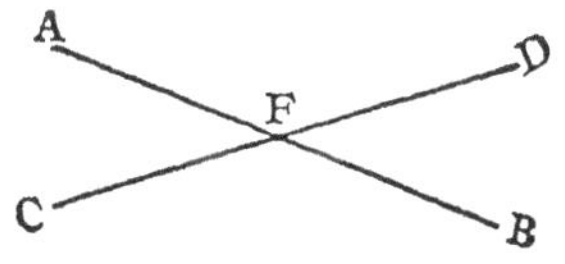

The demonstration is as follows: "For the angle CFA + the angle AFD = two right angles (prop. 13), and also the angle AFD + the angle DFB = two right angles; therefore the angle CFA + the angle AFD = the angle AFD + the angle DFB (axiom 1); and the common angle AFD being taken away from both, there remains the angle CFA = the angle DFB (axiom 3); but these are vertically opposite angles. In like manner it may be proved that the vertically opposite angles

AFD and BFC are equal." What are the concepts and propositions employed in the proof of this theorem? We have concepts of number (two, *e. g.*), straight lines, point, circle, intersection, angles, right angles, vertically opposite angles, equality, addition, subtraction, remainder, etc.

We employ, among the propositions, Proposition XIII (When a straight line standing upon another straight line makes angles with it they are either two right angles or together equal to two right angles); this proposition was proved from certain others all resting ultimately upon the axioms and postulates. We employ Postulate 3 (A circle may be described from any center, with any interval from that center); Postulate 1 (A straight line may be drawn from any one point to any other point); Definition 12 (A circle is a plane figure bounded by one line called the circumference or periphery; to which all straight lines drawn from a certain point within the figure, are equal); also Axiom 1 (Things which are equal to the same thing are equal to each other).

Geometrical demonstrations are made with reference to figures, and it might seem as if we were really dealing with a concrete particular case instead of with general concepts alone; but the concrete case is simply an illustration and the whole demonstration is absolutely general. Accurate measurements of the figure would undoubtedly show inequalities between the vertically opposite angles for the reason that the lines are not absolutely straight; the figure symbolizes the intersection of any two absolutely straight lines and the demonstration is true of all such lines and only of such.

Similar statements apply to all the figures used in demonstrating geometrical propositions. Euclidian geometry as a whole is simply a complex system built up in the manner illustrated by the example above.

The data of the geometrician are comparatively few and simple. His general principles are already determined; his tests of truth are agreement with his principles, and not absolute agreement with concrete facts; however, he believes that if it were possible to observe and measure the facts accurately, and if facts could be found to agree with his definitions of straight line, circle, etc., for example, his conclusions would be found in correspondence with them.[3]

The system which we have just examined aims at the organization of a set of judgments having a general application and not with any specific data. The science of geometry is not interested in this or that geometrical figure; it gives us information regarding figures of certain kinds, leaving the question of its application to particular cases to the applied sciences. All of the so-called "pure sciences" are like geometry in this respect. They deal with general principles, not with particular cases. Sometimes the arts and sciences are distinguished along this line, the arts being defined as the application in practice of the principles embodied in the sciences. The science of geometry is a system founded upon a certain few general principles which are described as axiomatic.

Most of the sciences include, besides axiomatic prin-

[3] Recently, other systems of geometry have been built up. The methods employed are the same as those of Euclidian geometry, but they assume different postulates. The only requirement is that the system built upon them shall be a coherent whole.

ciples, others which are the outcome of the application of the methods of science to the study of empirical data. The science of *Mechanics* is an example of this. As we have seen, it "lays down propositions which are true in the same way of all fluids, all gases, etc." These laws were discovered as a result of a multitude of observations of the behavior of particular bodies of fluid, gas, etc.; the observations were recorded, classified and made the basis of inductive inferences, which were tested as completely as possible. Hence all the processes included in the scientific method of establishing laws are employed, but strong emphasis is placed upon the establishment of further conclusions on the basis of these laws, in other words, the *deductive* part of scientific method is emphasized.

Somewhat similar statements may be made regarding the science of *Chemistry*, though chemistry is perhaps less independent of particular facts and less able to proceed deductively on the basis of generalizations already established; and certain departments of chemistry are concerned rather with the classification of facts than with attempts to found a deductive science. Observation, analysis, classification and induction are all employed. Still, even in chemistry there is much that is deductive, and the interest of the chemist in particular facts tends to become only indirect.

Biology and *Psychology* may also be mentioned at this point. Both are interested in concrete phenomena mainly for the sake of the general conclusions which may be based upon them. Both employ the whole of scientific method, including the use of *statistics, averages* and *probability*. Both are farther than chemistry from the point at which a science begins to be primarily deductive. The employment of statistics and of methods of exact measurement has become of very great importance in these two sciences within the past few years. Reference to the recent literature of both subjects makes this very evident. As examples in psychology we might cite studies of animal behavior and studies in child psychology;[4] in biology, the investigations connected with Mendel's Law, the study of Variation, and so on.[5]

The pages from the works of Professor James (pages 280-284), illustrate psychological analysis, which does not employ statistics; and the essay from Huxley's works (pages 287-300) illustrates the same thing in the field of biology. Neither passage can be regarded as representative of the studies which are most frequent at the present time. Both are exceptionally broad in scope and deductive in character, but both do illustrate the desire of scientists in both fields to arrive at general conclusions and to establish coherent systems.

Systems Which are More Concerned with Concrete Data.—We wish now to consider systems which have to do with concrete data in a different way; systems which give to specific instances their exact place in a system of concrete facts; systems which not only present a body of general laws, but also apply them to the explanation of specific cases; systems which state completely the causes of given phenomena or enable us to

[4] See Yerkes, *The Dancing Mouse;* Washburn, *The Animal Mind;* Thorndike, *Educational Psychology;* and articles in journals.

[5] See Punnett, *Mendelism;* and articles in journals.

establish the existence of given events or situations not at present open to observation.

History, in so far as it is concerned with the reconstruction of the past, is a case in point; so also is the criminal lawyer's attempt to discover the individual guilty of a crime; again *Geology*, when it aims at the discovery of past changes and conditions of the earth falls in the same class. In geology the primary data are almost entirely the present character of the earth's surface and the changes which are constantly going on. It includes, of course, many of the data and conclusions of physics, chemistry, and biology. On the basis of these data it is able to arrive at well-founded conclusions concerning changes which never could have been observed. It assumes, as does all science, that unless there is evidence to the contrary, the past and the unobserved are like the present and the observed.

"Many of the changes which have indisputably taken place are such as no man has ever observed, because they are brought about so slowly or so deep down within the crust that no direct observation is possible, and we can only infer the mode of procedure by examining the result. No human eye has ever witnessed the birth of a mountain range, or has seen the beds of solid rock folded and crumpled like so many sheets of paper, or observed the processes by which rock is changed in all its essential characteristics; 'metamorphosed' as it is technically called."[6] Conclusions of this character imply the existence of a well organized body of knowledge concerning the relations of facts in the given field.

[6] Scott, *Introduction to Geology*, p. 30.

The data of the science were obtained by observation: the facts observed were classified and correlated; laws were gradually discovered and verified; systems were constructed and rejected until eventually a set of general principles emerged upon which geologists could agree; but the working out of the details and the exposition of the system of concrete facts involved—in other words the history of the earth—has progressed only a little way. In geometry observation is only incidental; in any concrete science it is a constant necessity. In a historical science the particular fact occupies the center of the field; in other sciences the primary interest is in generalizations.

The discovery of the planet Neptune implied the construction of a similar system in *Astronomy*. The planets move in elliptical orbits; these orbits, however, are not perfect ellipses; there are variations (perturbations) due to the influence of other planets. The amount of variation due to any planet can be calculated. In the case of the planet Uranus, after all the perturbations due to the known planets had been taken into account, there remained a residue unaccounted for. Adams and Leverrier calculated that this residue could be accounted for by the hypothesis of another planet in a given direction and at a given distance from Uranus; and this planet was soon after discovered with the aid of the telescope and was named Neptune. The discovery of Neptune presupposed a knowledge of the general laws of the solar system and a comprehensive description of the relation in which its known members stood to each other. It was the result of the application of the general laws to a concrete situation.

Systems of Historical Facts.—Both astronomy and geology find a large part of the concrete data with which they deal open to observation; geology to some extent and astronomy to a great extent depend upon the testimony of past observers. In the courts and in the investigations of the historian, testimony is of the first importance. We may then classify the data in these investigations as follows:

1. Material facts.
2. Testimony.

Examples of material facts would be the articles found on the scene of the crime, etc., etc.; and in historical inquiries, the ruins of buildings, roads, and other public works, tombs, ancient implements, works of art, etc., etc. The testimony may be either oral or written. Usually the historian must rely principally upon written testimony. It is of many kinds, from the pictographs of primitive man to historical accounts like those of Thucydides and Tacitus. The problems of the historian are to collect the data, weigh the value of each of the items, and construct an account which will best organize the data into a coherent whole.

A. His first duty will be to *collect* all *the data* possible. The methods employed by the lawyer are sufficiently familiar in their general outlines. Details of his methods are beyond the scope of our discussion. We shall consider the procedure of the historian a little more closely. He must search for documents. "*Documents are traces which have been left by the thoughts and actions of men of former times,*" [7] whether

[7] Langlois and Seignobos, *Introduction to the Study of History*. This is a most valuable introduction to the method of history and should be read by every student of scientific method.

material facts, such as works of art and the like, or written records. Some historians have described events so recent that it was possible to obtain the testimony of eye-witnesses. They were thus able to obtain a quantity of testimony not available for the later writer and to cross-examine their witnesses. Usually the discovery of documents has meant a search in all sorts of places for the testimony of contemporaries; the expeditions and excavations carried on by the archæologists have as their object the discovery of such records. Archives, state papers, early histories, memoirs, inscriptions, etc., furnish the chief sources. The problem grows easier with the growth of collections and libraries. But new documents are constantly being found, and no one can say when the data have all been discovered.

B. After the collection of the data comes the *estimation of the value of its various items.* Estimating the value of *testimony* is a special problem and we shall treat of it first.[8] The problem is to discover what facts the documents establish. The questions to be raised are such as these: Who did such and such an act? Who wrote such and such a poem? Who founded Rome? The answer involves the establishment of a system of facts and inferences justifying some one conclusion to the exclusion of all others. The establishment of such a system means the *discovery of facts,* their *classification,* the *construction of hypotheses,* and

[8] Estimating the value of testimony is so important a matter in inquiries of the sort we are now considering that the questions involved will be discussed somewhat more fully than in the introductory chapter. All the machinery of scientific method may be needed to enable the historian to decide whether or not a record is trustworthy.

the *verification* of these, and the *organization* of facts and generalizations into a system. We know facts by observation, by memory aided by inference, by inference from the testimony of others, by inference from remains of former human activities, and from all sorts of natural events, and natural processes; we group the data with reference to their bearing on various parts of the problem; we use the inferences based upon past experience and formulate new ones; we test our constructions by all known means.

Sometimes observations of natural phenomena may make up most of the data. Observations made by others may sometimes be easily verified. Experiments may be repeated; the problem of getting correct descriptions of the facts may frequently be comparatively easy. But the sort of case which illustrates the establishment of a system of concrete facts in all its complexity is that in which *human testimony*, as well as all other kinds of data is employed to determine the existence and character of some fact. Let us examine in outline the processes involved in guarding against error in the use of testimony. A number of *preliminary questions* must be raised:

(1) The first of them is this: *What is the testimony? What does the witness say? What is the content of the document.* The testimony is the starting point. We must know what the words mean, what they purport to tell, what facts they are intended to represent. (*a*) In the case of the witness on the stand this question, though all-important, is usually comparatively easy of solution. In case of doubt he can be called upon for further statements, which will make plain his meaning. When he speaks in a foreign language there

is more difficulty, but in any ordinary description of facts, the difficulty is not great. (*b*) In the case of historical documents the difficulty may be insurmountable. For centuries hieroglyphics could not be interpreted at all; and translations from ancient documents are always attended with danger. The difficulty of finding in one language exact equivalents of the words of another needs no emphasis. In such a case the most thoroughgoing comparison of the two languages may be necessary in order to determine the meaning of a document. Whole sciences, such as epigraphy or palæography, are devoted to the interpretation of ancient writings.

(2) If the meaning of the statements contained in the testimony has been made clear, the next question is, of course: *Are the statements true?* And in order to answer this we may ask next: (*a*) Who made the statements, and is he qualified by knowledge, honesty and accuracy sufficient to enable us to rely upon what he says? In the law courts, *the identity of the witness* is the first thing to be determined and made a matter of record. When the witness is before us the question of his identity is usually very easy to answer, though there are, of course, numerous cases in which error and deception might occur. And in the case of the prisoner, the determination of identity is often an extremely difficult matter, involving testimony and many other kinds of evidence.

In the case of written documents the determination of authorship is one of the most difficult problems which the historian has to solve.[9] The name upon the title page of a book is, by itself, not conclusive as evidence

[9] See Langlois and Seignobos, Book II, chap. iii.

of authorship. In modern books the indications of authorship are usually fully given and are ordinarily reliable; fraud is possible, but is usually easily detected, though forgeries in the name of dead authors may be successful. But in the case of early books, and above all in the case of manuscripts, the difficulties are very great; in the first place there may be no formal indications of authorship; or the work is perhaps ascribed to such and such a person. Was he the author?

We should ask first: Did such a person ever exist? To answer this, only the testimony of contemporaries would be conclusive. In many cases it is not necessary to pause over this question, for there may be no doubt about the existence of the author; but in case there should be, the testimony to his having existed must be tested like any other such evidence. Consistency of testimony and corroboration of one piece of testimony by another are necessary here as elsewhere. One of the best known examples of this problem is that of the existence of Homer. Absence of testimony in this and other cases is usually a presumption against the truth of what is alleged.

Granted that there was such a person as the author, did he write the document before us? What evidence is there of the genuineness of the document? The evidence is of two kinds, *internal* and *external*. In examining *internal* evidence, the question is: Is the document such as the alleged author could or would have written? Is the handwriting of a sort that was employed during the lifetime and in the country, etc., of the supposed author? If the document is in handwriting of the Eleventh or the Thirteenth Century, it was not written by an author of the Twelfth Century. And with regard to the style and forms of expression the same question may be raised. In legal documents this is a valuable test, for legal phraseology is very definite. Modern words or phrases in a supposed ancient writing are, of course, conclusive against any argument for its genuineness.

Mention of facts and allusions to events of every sort are most valuable in this connection; and lastly, the opinions expressed or implied in the document are of great assistance in determining its genuineness, for some opinions could not possibly have been held at the time the document purports to have been written. *External evidence* is to be found in references to the document, quotations, etc., by contemporaries or by those of later periods.

Another complication is often present to add to the difficulty of determining authorship; the document may be the work of two or more individuals, or changes may have been introduced by those who have edited the texts, or there may have been mistakes in copying. In all these cases one should not necessarily reject the work as a whole; it may give some information and we may be able to detect the changes from the original, and it may be of great importance to determine just what was done by the original author or just what was written by each of several collaborators. The methods to be employed are of course those described above. It is simply a question of several authors instead of one.

(*b*) Having assured ourselves of the identity of the witness or witnesses, the next question is: *Was he in a position to know the facts to which he is testifying?* Does he assert that he witnessed them himself?[10] And was it humanly possible for any one to have observed the facts in question? And if this question can be answered in the affirmative, we have next to ask whether there is any reason why the *witness himself* could not have observed them. Was he (*i*) *competent* to observe and remember the facts? Was there any defect in his powers of observation, or is there any evidence against his having been at a place where the observation could have been made? Here again, when we have the witness before us, the difficulty of solving the problem is much easier than when we have to rely upon written testimony, or other evidence of an indirect sort.

Cross-examination is a method of getting at once an addition to the testimony on the points already raised and of furnishing immediately statements which will corroborate or disagree with statements already made, or with known facts. In Lincoln's first murder trial, the chief witness had testified to seeing the murder committed by the prisoner. In the cross-examination he added a number of details: that the shooting was at ten o'clock at night, in beech timber, in August, that he was twenty feet or more away, that he could see the pistol and how it hung; that the nearest lights were half a mile away, and that he saw it all by moonlight. Lincoln showed that the moon did not rise till one o'clock in the morning. Cross-examination may bring out inconsistencies due to dishonesty as well as incompetence, as shown in this example. Where cross-ex-

10 As Bain asserts (*Logic*, Appendix I), "The supreme canon of historical evidence is testimony of a contemporary"—of one who may have observed the fact.

amination is impossible, as in written testimony, it may be impossible to convict a dishonest witness.

(*ii*) If the witness has withstood all the preceding tests we have next to ask whether there is anything in *his record* which would lead us to doubt his honesty or whether he is likely to have any *reason for falsifying in the present case.* The two questions are distinct; a general good reputation would be a presumption in favor of his honesty in the present case, but it would not make it certain that he was proof against all temptation. In the courts, cross-examination of the witness himself and the testimony of other witnesses furnish the data for answering the question; in written testimony, other statements of his own and the statements of his contemporaries must be taken into account; what is implied is often more important than what is stated outright.

There are two cases in which the testimony of a witness may be regarded as particularly free from wilful falsification. The first is that in which the witness believes the evidence is to his discredit or disadvantage. One exception must be made: the witness, for the sake of satisfying a grudge or shielding someone else, might be willing to sacrifice his own reputation and advantage. In any case, such testimony would need further corroboration, but it would, with the exception above mentioned, be excellent evidence of the good faith of the witness.

The second case is that in which evidence is given undesignedly. The witness makes statements of whose import he is unaware or he is surprised into statements which bring out facts which he has been attempting to conceal. An incident related in Voltaire's "Zadig" will illustrate this:

> Zadig's master, Setoc, had lent money in the presence of two witnesses who had died before the debt was paid. The debtor denied having received any money. The money had been counted out upon a stone near Mount Horeb. Zadig undertook to conduct the case. He summoned the debtor before a tribunal and demanded that the five hundred ounces of silver be returned to his master. "Have you witnesses?" asked the judge. "No," replied Zadig, "they are dead; but there is a large stone on which the money was counted out; if it please Your Highness to order that the stone be sought out, I hope that it will bear witness; the debtor and I will remain here until the stone arrives; I will have it hunted up at the expense of my master." "Very well," replied the judge, and he turned his attention to something else. At the end of the sitting he said to Zadig, "Well, your stone has not yet arrived?" The debtor laughed and said: "If your Highness should remain here till to-morrow the stone would not have arrived; it is more than six miles away and it would require fifteen men to move it." "Well," cried Zadig, "I told you that the stone would bear witness; since this man knows where it is, he admits that it was upon it that the money was counted out."—Voltaire, *Zadig,* Chap. x.

Testimony which is false is of course evidence of something,—of the opinion of the witness, or of his character, or of the existence of certain ideas, etc., at the time in which he lived.

But all these questions are more or less preliminary. A witness of good reputation and a good observer, with no motive to falsify, may, of course, be mistaken in the case under examination. And a witness who is not usually reliable or who is a bad observer or one with every reason to falsify, may be telling the truth. In the first case the presumption would of course be in favor of the testimony and in the second against it, but in neither case could we regard the reliability of the testimony as settled. There are *three conditions to the acceptance of every piece of testimony.* (1) It must be self-consistent and internally coherent; (2) it must be consistent and coherent with other known facts relating to the same case; (3) it must be consistent with ordinary experience.

1. If a witness, in one part of his testimony, makes a statement inconsistent with what he has stated previously, his testimony is discredited. One of his statements may be true or both may be false; it may be possible to show that one of them is consistent with the rest of his testimony, while the other is not, but the disagreement of these statements would tend to throw doubt upon the others, and without external corroboration his whole story would be open to question. Moreover, his statements, to have the greatest force, must not only be consistent, they must be coherent; they must describe a connected series of events. If a lawyer can by cross-examination show that the witness has made inconsistent statements, the force of his testimony is often entirely destroyed; and it is usually very much weakened, at the very least.

But even the most coherent body of testimony would, by itself, be insufficient to prove the existence of the facts alleged. Otherwise we should be obliged to accept as true many acknowledged fictions. Indeed, too great coherence, too good a story, rouses the suspicion that it has been manufactured or at least modified, since most men are too inaccurate both in observing and in remembering to describe any complex set of events without minor inconsistencies.

By itself, then, internal evidence is not conclusive; without the support of other testimony or of facts otherwise known any piece of testimony must be held as doubtful. A possible exception might be noted: if the testimony were of such a character that its falsity would be more difficult to understand and explain than its truth, we should have some ground for accepting it even in the absence of other corroboration; but such cases would obviously be rare.

Negative Evidence.—The absence of testimony to the existence of a fact which could hardly have failed of mention by contemporaries is a strong presumption against its existence. If an alleged work, or doctrine, or what not, is referred to by no contemporary and is first mentioned by some later writer, it is probably false. The tradition regarding individuals, cities, and so on, often becomes more extensive and circumstantial as the objects of the tradition get farther away.[11]

2. In oral testimony, one of the most frequent sources of corroboration is to be found in the testimony of other witnesses. But too close an agreement, instead of being evidence of the truth of the testimony, raises a suspicion that the witnesses are in collusion and that the whole story may be false. The inevitable inaccuracies of observation and of memory render it practically impossible that two witnesses should tell stories that should agree in all particulars; some differences are to be expected, and sometimes there are differences regarding some of the most important points. In writ-

11 For illustrations, see Hayward, *Essays*, "The Pearls and Mock Pearls of History."

ten testimony, if there are several documents very closely similar, the presumption is that, instead of being independent pieces of testimony, they are all derived from the same source. This presumption is particularly strong if the errors happen to be the same in all; if, for example, the same words are misspelled or the same misstatements of fact are made in all, this is good evidence that one of the documents was the source of the rest or that all were derived from some common source. Of course disagreement does not prove the truth of any one of the bodies of testimony, but it is good evidence of their independent origin. Where the truth lies must be discovered by further comparison and construction.

3. The third test mentioned above was *agreement with ordinary experience.* What is known as the Argument to Antecedent Probability would be included here. We ask: Is the alleged event one that would have been probable in the circumstances? Is it consistent with the known laws of nature and their familiar modes of operation? This test might, in practice, be applied first; if, for example, testimony contained statements violating all ordinary experience or well-tested laws of nature, we might decline to go to the trouble of applying the other tests. But this test is not necessarily final, for statements which disagree with our past experience are very frequently found to be true and it has more than once happened that a supposed law of nature has been stated too broadly and has needed qualification. A too ready rejection of unusual statements is no more justified by a sound method than is a too ready credulity. But if the application of other tests leaves the truth of the testimony inconclusive, a viola-

tion of ordinary experience would warrant the rejection of the testimony. If alleged facts were in contradiction to supposed laws of Nature, their existence could be established only by evidence which was stronger than the whole body of evidence in favor of the supposed law. Proof of the violation of all the laws of Nature would be impossible; for all proof requires the use of some law. We are often over-hasty in concluding that a statement is inconsistent with another or with some of the consequences of the latter, and we tend too readily to condemn anything which is apparently inconsistent with established principles. Rejection of the Copernican Hypothesis on the ground of its inconsistency with the principles of religion was a case of this sort.

There are two or three topics which call for a little further discussion at this point. One of these is what is known as *hearsay evidence;* in this the witness reports not what he himself observed, but what he has heard some one else describe. Its value is very much less than is that of testimony to the fact itself. To the errors of the original observer, the errors of observation, of memory, of description, and possible bad faith on the part of the original observer, we have added those of a second person, who is liable to the same errors and defects with regard to the *words* of the first.

Tradition is simply hearsay evidence with a multiplication of the number of intermediates between the last hearer and the original observer, if indeed, there were any observations at the beginning. It is evident that tradition is exceedingly poor evidence of the existence of any fact.

Circumstantial evidence is merely indirect evidence; there may be no witnesses who have observed the facts themselves, but certain other facts may be incapable of explanation on any theory other than the one which asserts the existence of these facts.

We have been discussing the problems of discovering and evaluating historical facts; the principles involved in building up a system when the data are derived from both testimony and observation are, of course, the same as those involved in constructing a system from data obtained in any other way. We wish to get a complete, coherent whole. The framework of general statements or laws, and the particular structure which we build, must provide a place for the facts which are the materials to be built into a system. If there are facts which do not fit, if there are parts of the framework which interfere with each other, if the laws of such structures are violated, the construction can not be accepted. A place for every fact, and a complete structure when all the facts are in place, are the requirements of a scientific structure, or, in other words, of any structure of knowledge which is to satisfy the demands of a reasonable being.

The details of historical construction are beyond the scope of this outline and an illustration which should show them with any degree of fullness would occupy too much space. Huxley's argument, quoted on pages 287-300), illustrates a few of the points. The student is referred to Langlois and Seignobos's *Introduction to the Study of History* for further discussion and illustration.

EXERCISES

In the following examples, outline the argument and state the general principles employed, and the way in which these principles were or might have been established; examine each step in the reasoning and determine whether or not it is valid; where inductive inference is used describe its foundation, and criticise, if possible; where the argument is incomplete state what would be necessary in order to complete it:

I. Professor James's argument for his theory of the emotions, as given in his *Psychology, Briefer Course,* pp. 375 to 380.

"The feeling, in the coarser emotions, results from the bodily expression. Our natural way of thinking about the coarser emotions is that the mental perception of some fact excites the mental affection called the emotion, and that the latter state of mind gives rise to the bodily expression. My theory, on the contrary, is that *the bodily changes follow directly the perception of the exciting fact, and that our feeling of the same changes as they occur IS the emotion.* Common sense says, we lose our fortune, are sorry and weep; we meet a bear, are frightened and run; we are insulted by a rival, are angry and strike. The hypothesis here to be defended says that this order of sequence is incorrect, that the one mental state is not immediately induced by the other, that the bodily manifestations must first be interposed between, and that the more rational statement is that we feel sorry because we cry, angry because we strike, afraid because we tremble, and not that we cry, strike, or tremble because we are sorry, angry, or fearful, as the case may be. Without the bodily states following on the perception, the latter would be purely cognitive in form, pale, colorless, destitute of emotional warmth. We might then see the bear and judge it best to run, receive the insult and deem it right to strike, but we should not *feel* afraid or angry.

"Stated in this crude way, the hypothesis is pretty sure

to meet with immediate disbelief. And yet neither many nor far-fetched considerations are required to mitigate its paradoxical character, and possibly to produce conviction of its truth.

"1. To begin with, *particular perceptions certainly do produce widespread bodily effects by a sort of immediate physical influence, antecedent to the arousal of an emotion or emotional idea.* In listening to poetry, drama, or heroic narrative, we are often surprised at the cutaneous shiver which like a sudden wave flows over us, and at the heart-swelling and lachrymal effusion that unexpectedly catch us at intervals. In hearing music the same is even more strikingly true. If we abruptly see a dark moving form in the woods, our heart stops beating, and we catch our breath instantly and before any articulate idea of danger can arise. If our friend goes near to the edge of a precipice, we get the well-known feeling of 'all-overishness,' and we shrink back, although we positively *know* him to be safe, and have no distinct imagination of his fall. The writer well remembers his astonishment, when a boy of seven or eight, at fainting when a horse was bled. The blood was in a bucket, with a stick in it, and, if memory does not deceive him, he stirred it round and saw it drip from the stick with no feeling save that of childish curiosity. Suddenly the world grew black before his eyes, his ears began to buzz, and he knew no more. He had never heard of the sight of blood producing faintness or sickness, and he had so little repugnance to it, and so little apprehension of any other sort of danger from it, that even at that tender age, as he well remembers, he could not help wondering how the mere physical presence of a pailful of crimson fluid could occasion in him such formidable bodily effects.

"2. The best proof that the immediate cause of emotion is a physical effect on the nerves is furnished by *those pathological cases in which the emotion is objectless.* One of the chief merits, in fact, of the view which I propose, seems to be that we can so easily formulate by its means pathological cases and normal cases under a common scheme. In every asylum we find examples of absolutely unmotived fear, anger, melancholy, or conceit; and others of an equally unmotived apathy which persists in spite of the best outward reasons why it should give way. In the former cases we must suppose the nervous machinery to be so 'labile' in some one emotional direction that almost any stimulus (however inappropriate) causes it to upset in that way, and to engender the particular complex of feelings of which the psychic body of the emotion corresponds. Thus, to take one special instance, if inability to draw a deep breath, fluttering of the heart, and that peculiar gastric change felt as 'precordial anxiety,' with an irresistible tendency to take a somewhat crouching attitude and to sit still, with perhaps other visceral processes not now known, all spontaneously occur together in a certain person, his feeling of the combination *is* the emotion of dread, and he is the victim of what is known as morbid fear. A friend who has occasional attacks of this most distressing of all maladies tells me that in his case the whole drama seems to center about the region of the heart and respiratory apparatus, that his main effort during the attacks is to get control of his inspirations and to slow his heart, and that the moment he attains to breathing deeply and holding himself erect, the dread, *ipso facto,* seems to depart.

" The emotion here is nothing but the feeling of a bodily state, and it has a purely bodily cause.

" 3. The next thing to be noticed is this, that *every one of the bodily changes, whatsoever it be, is FELT, acutely or obscurely, the moment it occurs.* If the reader has never paid attention to this matter he will be both interested and astonished to learn how many different bodily feelings he can detect in himself as characteristic of his various emotional moods. It would be perhaps too much to expect of him to arrest the tide of any strong gust of passion for the sake of any such curious analysis as this; but he can observe more tranquil states, and that may be assumed to be true of the greatest which is shown to be true of the less. Our whole cubic capacity is sensibly alive; and each morsel of it contributes its pulsations of feeling, dim or sharp, pleasant, painful, or dubious, to that sense of personality that every one of us unfailingly carries with him. It is surprising what little items give accent to these complexes of sensibility. When worried by any slight trouble, one may find that the focus of one's bodily consciousness is the contraction, often quite inconsiderable, of the eyes and brows. When momentarily embarrassed, it is something in the pharynx that compels either a swallowing, a clearing of the throat, or a slight cough; and so on for as many more instances as might be named. The various permutations of which these organic changes are susceptible make it abstractly possible that no shade of emotion should be without a bodily reverberation as unique, when taken in its totality, as is the mental mood itself. The immense number of parts modified is what makes it so difficult for us to reproduce in cold blood the total and integral expression of any one emotion. We may catch the trick with the voluntary muscles, but fail with the skin, glands, heart, and other viscera. Just as an artificially imitated sneeze lacks something of the reality, so the attempt to imitate grief or enthusiasm in the absence of its normal instigating cause is apt to be rather 'hollow.'

"4. I now proceed to the vital point of my whole theory, which is this: *If we fancy some strong emotion, and then try to abstract from our consciousness of it all the feelings of its bodily symptoms, we find we have nothing left behind,* no 'mind stuff' out of which the emotion can be constituted, and that a cold and neutral state of intellectual perception is all that remains. It is true that, although most people, when asked, say that their introspection verifies this statement, some persist in saying that theirs does not. Many of them cannot be made to understand the question. When you beg them to imagine away every feeling of laughter and of the tendency to laugh from their consciousness of the ludicrousness of the object, and then to tell you what the feeling of its ludicrousness is like, whether it be anything more than the perception that the object belongs to the class 'funny,' they persist in replying that the thing is a physical impossibility, and they always must laugh if they see a funny object. Of course the task proposed is not the impossible one of seeing a ludicrous object and annihilating one's tendency to laugh. It is the purely speculative one of subtracting certain elements of feeling from an emotional state supposed to exist in its fullness, and saying what the residual elements are. I cannot help thinking that all who rightly apprehend this problem will agree with the proposition above laid down. What kind of an emotion of fear would be left if the feeling neither of quickened heart-beats nor of shallow breathing, neither of trembling lips nor of weakened limbs, neither of goose-flesh nor of visceral stirrings, were present, it is quite impossible for me to think. Can one fancy the state of rage and picture no ebullition in the chest, no flushing of the face, no dilatation of the nostrils, no clenching of the teeth, no impulse to vigorous action, but in their stead limp muscles, calm breathing, and a placid face? The present writer, for one, certainly cannot. The rage is as completely evaporated as the sensation of its so-called manifestations, and the only thing that can possibly be supposed to take its place is some cold-blooded and dispassionate judicial sentence, confined entirely to the intellectual realm, to the effect that a certain person or persons merit chastisement for their sins. In like manner of grief: what would it be without its tears, its sobs, its suffocation of the heart, its pang in the breast bone? A feelingless cognition that certain circumstances are deplorable and nothing more. Every passion in turn tells the same story. A disembodied human emotion is a sheer nonentity. I do not say that it is a contradiction in the nature of things, or that pure spirits are necessarily condemned to cold, intellectual lives; but I say that for *us* emotion disassociated from all bodily feeling is inconceivable. The more closely I scrutinize my states, the more persuaded I become that whatever 'coarse' affections and passions I have are in very truth constituted by, and made up of, those bodily changes which we ordinarily call their expression or consequence; and the more it seems to me that, if I were to become corporeally anæsthetic, I should be excluded from the life of the affections, harsh and tender alike, and drag out an existence of merely cognitive or intellectual form."

II. Extract from a lecture by A. H. Fison on "The Evolution of Double Stars" in *Lectures on the Method of Science,* Edited by T. B. Strong:

The leading points of Darwin's [12] investigations of the

[12] Professor George Darwin.

past history of the moon: " From the fact that the intensity of the moon's attraction is greater upon the parts of the Earth that are nearer to it than upon the parts that are more remote there arises a tendency for the earth to become stretched along the diameter that is at that particular instant directed towards the Moon. If the Earth were fluid, it would yield to this tendency, but, as it is in the main solid, it is unable to do so. The waters upon its surface are, however, free, and they consequently flow, continually tending to accumulate in two high tides, one immediately under the Moon, and the other at the part of the Earth's surface that is most remote from it. If the period of the Earth's rotation was the same as that of the Moon's revolution round it, the Moon would continually face the same regions of the Earth, and in the course of time, possibly a few months or years, the water would reach a position of equilibrium, forming permanent high tides at the opposite ends of the diameter that would then be permanently directed to the Moon.

" This simple condition is, however, profoundly modified by the Earth's rotation. As the Earth turns under the Moon in a period of slightly less than twenty-five hours, the regions presented to the Moon—those at which the water tends to accumulate—are continually changing, and before any portion of water could move appreciably towards them, the forces acting upon it would change it and it would be urged in some other direction. The problem thus becomes extremely complicated. The general result, however, is, that in its continual endeavor to move towards the ends of the terrestrial diameter that is at each given instant pointing to the Moon, the water on the Earth's surface is thrown into the continual motion that we recognize as tidal ebb and flow.

" If the movement of the water were unresisted by friction, tidal ebb and flow would possess no cosmical significance, but friction is experienced in the motion of the tidal wave over the surface of shores and estuaries, and in internal motions of the water itself. The destruction of motion by friction develops heat, and the Earth is consequently warmed by its tides; moreover, since heat is a form of energy, some other form of energy, equivalent in amount, must disappear in producing it. From considerations of a not very difficult nature, it can be shown that this energy is that of the Earth's rotation, so that we are presented with the remarkable fact that the speed of the Earth's rotation is being reduced in consequence of the tides. The period of the Earth's rotation determines the day, and consequently the day must be increasing in length. No doubt the rate of increase is now very slight, but there can be little doubt that this has not always been the case. The Earth was at one time a mass of molten rock, in which bodily tides must have been formed, while friction must have been far greater in the case of such a viscous mass than in water. Further, as we shall see, the Moon must have been nearer the Earth than it is now, and its tide-producing power consequently more intense. Under these conditions we can well imagine that the loss of rotation proceeded at a comparatively rapid pace, and that the day was formerly far shorter than it is at the present time.

"The slackening of the Earth's rotation is not, however, the only result of tidal friction. A reaction upon the Moon is inevitable, and it appears that, as a necessary consequence, the Moon must recede from the Earth, its orbital speed decreasing at the same time. Its period of revolution round the Earth, which we may define as the month, is therefore increasing, so that in consequence of the tides, the day and the month are both becoming longer. It follows, however, from simple considerations, that this cannot continue indefinitely. The day is increasing, and so also is the month, but there must come a time when the day must increase more rapidly than the month (already past), and it must ultimately overtake it. The length of each will then be fifty-five of our present days. The Earth, then rotating in the same period as that of the Moon's revolution round it, will continually present the same regions to the Moon, as the Moon already presents the same face to it. At the ends of the terrestrial diameter that will then be constantly pointing towards the Moon, permanent high tides will accumulate; ebb and flow, and with it tidal friction will cease, and a state of stable equilibrium will be reached. It is impossible to determine the epoch of this stage, but under the most favorable conditions it must be measured by hundreds of millions of years from the present time. . . . In the past the Earth must have rotated more rapidly, the Moon must have been nearer, and it must have revolved in a shorter period than at present. From the application of mathematics to the problem, Darwin has shown that there must have been a time when the Moon was quite close to the surface of the Earth, and, when in this condition, the further suggestive fact appears that its period of revolution, the month, coincided, as it will again coincide in the last stage, with that of the Earth's rotation, the day. Both must then have been between three and five hours in length. In this first, as in the last condition, we see Earth and Moon rotating as a whole about their common center of mass, each continually presenting the same face to each other. While, however, in the first condition they are nearly in contact and represent a passing phase, in the last they are far apart, and their condition is permanent. It is possible to show that the first stage could not have occurred less than 50,000,000 years ago."

III. Huxley's lecture on "The Demonstrative Evidence of Evolution" (with a few omissions):

"The occurrence of historical facts is said to be demonstrated, when the evidence that they happened is of such a character as to render the assumption that they did not happen in the highest degree improbable; and the question I now have to deal with is, whether evidence in favor of the evolution of animals of this degree of cogency is, or is not, obtainable from the record of the succession of living forms which is presented to us by fossil remains.

"Those who have attended to the progress of palæontology are aware that evidence of the character which I have defined has been produced in considerable and continually increasing quantity during the last few years. Indeed, the amount and the satisfactory nature of that evidence are somewhat surprising, when we consider the conditions under which alone we can hope to obtain it.

"It is obviously useless to seek for such evidence, except in localities in which the physical conditions have been such as to permit of the deposit of an unbroken, or but rarely interrupted, series of strata through a long period of time; in which the group of animals to be investigated has existed in such abundance as to furnish the requisite supply of remains; and in which, finally, the materials composing the strata are such as to insure the preservation of these remains in a tolerably perfect and undisturbed state.

"It so happens that the case which, at present, most nearly fulfills all these conditions is that of the series of extinct animals which culminates in the Horses; by which term I mean to denote not merely the domestic animals with which we are so well acquainted, but their allies, the ass, zebra, quagga, and the like. In short, I use 'horses' as the equivalent of the technical term *Equidæ,* which is applied to the whole group of existing equine animals.

"The horse is in many ways a remarkable animal; not least so in the fact that it presents us with an example of one of the most perfect pieces of machinery in the living world. In truth, among the works of human ingenuity it cannot be said that there is any locomotive so perfectly adapted to its purposes, doing so much work with so small a quantity of fuel, as this machine of nature's manufacture —the horse. . . . Look at the perfect balance of his form, and the rhythm and force of its action. The locomotive machinery is, as you are aware, resident in its slender fore and hind limbs; they are flexible and elastic levers, capable of being moved by very powerful muscles; and, in order to supply the engines which work these levers with the force which they expend, the horse is provided with a very perfect apparatus for grinding its food and extracting therefrom the requisite fuel.

"Without attempting to take you very far into the region of osteological detail, I must nevertheless trouble you with some statements respecting the anatomical structure of the horse; and, more especially, will it be needful to obtain a general conception of the structure of its fore and hind limbs, and of its teeth. But I shall only touch upon those points which are absolutely essential to our inquiry.

"Let us turn in the first place to the fore-limb. In

most quadrupeds, as in ourselves, the fore-arm contains distinct bones, called the radius and the ulna. The corresponding region in the horse seems at first to possess but one bone. Careful observation, however, enables us to distinguish in this bone a part which clearly answers to the upper end of the ulna. This is closely united with the chief mass of the bone which represents the radius, and runs out into a slender shaft which may be traced for some distance downwards upon the back of the radius, and then in most cases thins out and vanishes. It takes still more trouble to make sure of what is nevertheless the fact, that a small part of the lower end of the bone of a horse's fore-arm, which is only distinct in a very young foal, is really the lower extremity of the ulna.

"What is commonly called the knee of a horse is its wrist. The 'cannon bone' answers to the middle bone of the five metacarpal bones, which support the palm of the hand in ourselves. The 'pastern,' 'coronary,' and 'coffin' bones of veterinarians answer to the joints of our middle fingers, while the hoof is simply a greatly enlarged and thickened nail. But if what lies below the horse's 'knee' thus corresponds to the middle finger in ourselves, what has become of the four other fingers or digits? We find in the places of the second and fourth digits only two slender, splint-like bones, about two-thirds as long as the cannon bone, which gradually taper to the lower ends and bear no finger joints, or as they are termed, phalanges. Sometimes, small bony or gristly nodules are to be found at the bases of these two metacarpal splints, and it is probable that these represent rudiments of the first and fifth toes. Thus the part of the horse's skeleton which corresponds with that of the human hand contains one overgrown middle digit, and at least two imperfect lateral digits; and these answer, respectively, to the third, the second, and the fourth fingers in man.

"Corresponding modifications are found in the hind limb. In ourselves, and in most quadrupeds, the leg contains two distinct bones—a large bone, the tibia, and a smaller and more slender bone, the fibular. But, in the horse, the fibular seems, at first, to be reduced to its upper end; a short, slender bone, united with the tibia and ending in a point below, occupying its place. Examination of the lower end of a young foal's shin-bone, however, shows a distinct portion of osseous matter which is the lower end of the fibula; so that the apparently single lower end of the shin-bone is really made up of the coalesced ends of the tibia and fibula, just as the apparently single lower end of the fore-arm bone is composed of the coalesced radius and ulna.

"The heel of the horse is the part commonly known as the hock. The hinder cannon bone answers to the middle metatarsal bone of the human foot; the pastern, coronary, and coffin bones, to the middle toe bones; the hind hoof to the nail; as in the fore-foot. And, as in the fore-foot, there are merely two splints to represent the second and the fourth toes. Sometimes a rudiment of the fifth toe appears to be traceable.

"The teeth of the horse are not less peculiar than its limbs. The living engine, like all others, must be well stoked if it is to do its work; and the horse, if it is to make good its wear and tear, and to exert the enormous amount of force required for its propulsion, must be well and rapidly fed. To this end, good cutting instruments and powerful and lasting crushers are needful. Accordingly, the twelve cutting teeth of a horse are close-set and concentrated in the fore part of its mouth, like so many adzes or chisels. The grinders or molars are large, and have an extremely complicated structure, being composed of a number of different substances of unequal hardness. The consequence of this is that they wear away at different rates; and, hence, the surface of each grinder is always as uneven as that of a good mill-stone.

"I have said that the structure of the grinding teeth is very complicated, the harder and the softer parts being, as it were, interlaced with one another. The result of this is that, as the tooth wears, the crown presents a peculiar pattern, the nature of which is not very easily deciphered at first, but which it is important that we should understand clearly. Each grinding tooth of the upper jaw has an *outer wall* so shaped that, on the worn crown, it exhibits the form of two crescents, one in front and one behind, with their concave sides turned outwards. From the inner sides of the front crescent, a crescentic *front ridge* passes inwards and backwards, and its inner face enlarges into a strong logitudinal fold or *pillar*. From the front part of the hinder crescent, a *back-ridge* takes a like direction, and also has its *pillar*.

"The deep interspaces or *valleys* between these ridges and the outer wall are filled by bony substance, which is called *cement*, and coats the whole tooth.

"The pattern of the worn face of each grinding tooth of the lower jaw is quite different. It appears to be formed of two crescent-shaped ridges, the convexities of which are turned outwards. The free extremity of each crescent has a *pillar*, and there is a large double *pillar* where the two crescents meet. The whole structure is, as it were, embedded in cement, which fills up the valleys, as in the upper grinders.

"If the grinding faces of an upper and of a lower molar are applied together, it will be seen that the opposed ridges are nowhere parallel, but that they frequently cross; and that thus, in the act of mastication, a hard surface in the one is constantly applied to a soft surface in the other and vice versa. They thus constitute a grinding apparatus of great efficiency, and one which is repaired as fast as it wears, owing to the long continued growth of the teeth.

"Some other peculiarities of the dentition of the horse must be noticed, as they bear upon what I shall have to say by-and-by. Thus, the crowns of the cutting teeth have a peculiar deep pit, which gives rise to the well-known 'mark' of the horse. There is a large space between the outer incisors and the front grinder. In this space the adult male horse presents, near the incisors, one on each side, above and below, a caninc or 'tush,' which is commonly absent in mares. In a young horse, moreover, there is not infrequently to be seen, in front of the first grinder, a very small tooth, which soon falls out. If this small tooth be counted as one, it will be found that there are seven teeth behind the canine on each side, namely, the small tooth in question, and six great grinders, among which, by an unusual peculiarity, the foremost tooth is rather larger than those which follow it.

"I have now enumerated those characteristic structures of the horse which are of most importance for the purpose we have in view.

"To any one who is acquainted with the morphology of vertebrated animals, they show that the horse deviates widely from the general structure of mammals; and that the horse type is, in many respects, an extreme modification of the general mammalian plan. The least modified mammals, in fact, have the radius and ulna, the tibia and fibula, distinct and separate. They have five distinct and complete digits on each foot, and no one of these digits is very much larger than the rest. Moreover, in the least modified mammals, the total number of the teeth is very generally forty-four, while in horses the usual number is forty, and in the absence of the canines it may be reduced to thirty-six; the incisor teeth are devoid of the fold seen in those of the horse; the grinders regularly diminish in size from the middle of the series to its front end; while their crowns are short, early attain their full length, and exhibit simple ridges or tubercles, in place of the complex foldings of the horse's grinders.

"Hence, the general principles of the hypothesis of evolution lead to the conclusion that the horse must have been derived from some quadruped which possessed five complete digits on each foot; which had the bones of the fore-arm and of the leg complete and separate; and which possessed forty-four teeth, among which the crowns of the incisors and grinders had a simple structure, while the latter gradually increased in size from before backwards, at any rate in the anterior part of the series, and had short crowns.

"And if the horse has been thus evolved, and the remains of the different stages of its evolution have been preserved, they ought to present us with a series of forms in which the number of the digits becomes reduced; the bones of the fore-arm and leg gradually take on the equine condition; and the form and arrangement of the teeth successively approximate to those which obtain in existing horses.

"Let us turn to the facts and see how far they fulfil these requirements of the doctrine of evolution.

"In Europe abundant remains of horses are found in

the Quaternary and later Tertiary strata as far as the Pliocene formation. But these horses, which are so common in the cave deposits and in the gravels of Europe, are in all essential respects like existing horses. And that is true of all the horses of the latter part of the Pliocene epoch. But in deposits which belong to the earlier Pliocene and later Miocene epochs, and which occur in Britain, in France, in Germany, in Greece, in India, we find animals which are extremely like horses—which, in fact, are so similar to horses, that you may follow descriptions given in works upon the anatomy of the horse upon the skeletons of these animals—but which differ in some important particulars. For example, the structure of their fore and hind limbs is somewhat different. The bones which, in the horse, are represented by two splints, imperfect below, are as long as the metacarpal and metatarsal bones; and attached to the extremity of each is a digit with three joints of the same general character as those of the middle digit, only very much smaller. These small digits are so disposed that they could have had but very little functional importance, and they must have been rather of the nature of the dew-claws, such as are to be found in many animals. The *Hipparion,* as the extinct European three-toed horse is called, in fact, presents a foot similar to that of the American *Protophippus* (Fig. 12), except that, in the *Hipparion,* the smaller digits are situated further back, and are of smaller proportional size, than in the *Protohippus.*

"The ulna is slightly more distinct than in the horse; and the whole length of it, as a very slender shaft, intimately united with the radius, is completely traceable. The fibula appears to be in the same condition as in the horse. The teeth of the *Hipparion* are essentially similar to those of the horse, but the pattern of the grinders is in some respects a little more complex, and there is a depression on the face of the skull in front of the orbit which is not seen in existing horses.

"In the earlier Miocene and perhaps the later Eocene deposits of some parts of Europe, another extinct animal has been discovered, which Cuvier, who first described some fragments of it, considered to be a *Palæotherium.* But as further discoveries threw no light upon its structure, it was recognized as a distinct genus, under the name of *Anchitherium*.

"In its general characters, the skeleton of *Anchitherium* is very similar to that of the horse. In fact, Lartet and De Blainville called it *Palæotherium equinum* or *hippoides;* and De Christol, in 1847, said that it differed from *Hipparion* in little more than the characters of its teeth, and gave it the name of *Hipparitherium*. Each foot possesses three complete toes; while the lateral toes are much larger in proportion to the middle toe than in *Hipparion,* and doubtless rested on the ground in ordinary locomotion.

"The ulna is complete and quite distinct from the radius, though firmly united with the latter. The fibula seems also to have been complete. Its lower end, though intimately united with that of the tibia, is clearly marked off from the latter bone.

"There are forty-four teeth. The incisors have no strong pit. The canines seem to be well developed in both sexes. The first of the seven grinders, which, as I have said, is frequently absent, and, when it does exist, is small in the horse, is a good-sized and permanent tooth, while the grinder which follows it is but little larger than the hinder one. The crowns of the grinders are short, and though the fundamental pattern of the horse-tooth is discernible, the front and back ridges are less curved, the accessory pillars are wanting, and the valleys, much shallower, are not filled up with cement.

"Seven years ago, when I happened to be looking critically into the bearing of palæontological facts upon the doctrine of evolution, it appeared to me that the *Anchitherium,* the *Hipparion,* and the modern horse constitute a series in which the modifications of structure coincide with the order of chronological occurrence, in the manner in which they must coincide if the modern horses really are the result of a gradual metamorphosis, in the course of the Tertiary epoch, of a less specialized ancestral form. And I found by correspondence with the late eminent French anatomist and paleontologist, M. Lartet, that he had arrived at the same conclusion from the same data.

"That the *Anchitherium* type had become metamorphosed into the *Hipparion* type, and the latter into the *Equine* type, in the course of that period of time which is represented by the latter half of the Tertiary deposits, seemed to me to be the only explanation of the facts for which there was even a shadow of probability.

"And, hence, I have ever since held that these facts afford evidence of the occurrence of evolution, which, in the sense already defined, may be termed demonstrative.

"All who have occupied themselves with the structure of *Anchitherium,* from Cuvier onwards, have acknowledged its many points of likeness to a well-known genus of extinct Eocene mammals, *Palæotherium.* Indeed, as we have seen, Cuvier regarded his remains of *Anchitherium* as those of a species of *Palæotherium.* Hence, in attempting to trace the pedigree of the horse beyond the Miocene epoch and the Anchitheroid form, I naturally sought among the various species of Palæotheroid animals for its nearest ally, and I was led to conclude that the *Palæotherium minus* (*Plagiolophus*) represented the next step more nearly than any other form then known.

"I think that this opinion was fully justifiable; but the progress of investigation has thrown an unexpected light on the question, and has brought us much nearer than could have been anticipated to a knowledge of the true series of the progenitors of the horse.

"You are all aware that when your country was first discovered by Europeans, there were no traces of the existence of the horse in any part of the American continent. The accounts of the conquest of Mexico dwell upon the astonishment of the natives of that country when they first became acquainted with that astounding phenomenon—a man seated upon a horse. Nevertheless, the investigations of American geologists have proved that the remains of horses occur in the most superficial deposits of both North and South America, just as they do in Europe. Therefore, for some reason or other,—no feasible suggestion on that subject, so far as I know, has been made,—the horse must have died out in this continent at some period preceding the discovery of America. Of late years there has been discovered in your Western Territories that marvellous accumulation of deposits, admirably adapted for the preservation of organic remains, to which I referred the other evening, and which furnishes us with a consecutive series of records of the fauna of the older half of the Tertiary epoch, for which we have no parallel in Europe. They have yielded fossils in an excellent state of conservation and in unexampled number and variety. The researches

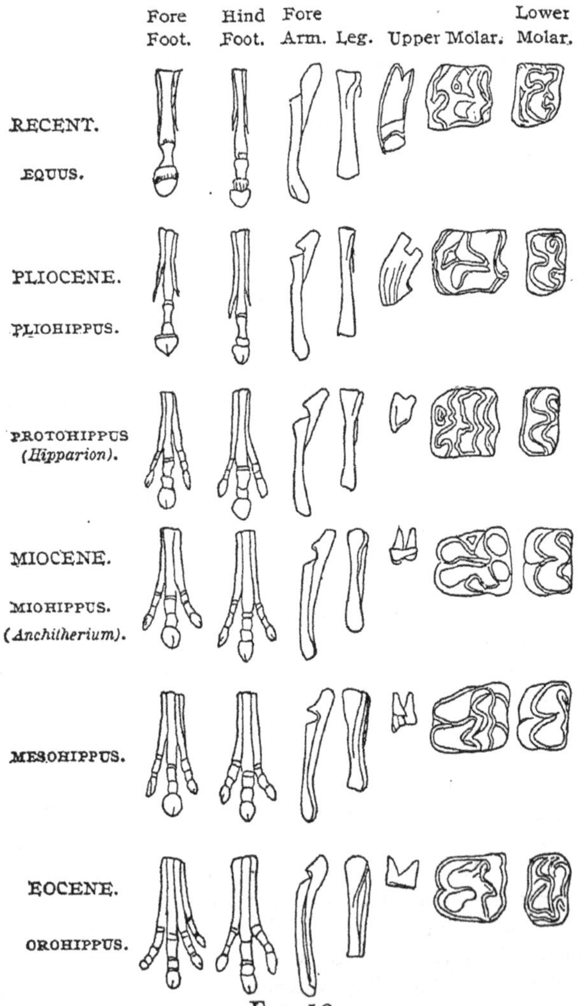

Fig. 12.

of Leidy and others have shown that forms allied to the *Hipparion* and the *Anchitherium* are to be found among these remains. And it is only recently that the admirably conceived and most thoroughly and patiently worked-out investigations of Professor Marsh have given us a just idea of the vast fossil wealth and of the scientific importance of these deposits. I have had the advantage of glancing over the collections in Yale Museum, and I can truly say that, so far as my knowledge extends, there is no collection from any one region and series of strata comparable for extent, or for the care with which the remains have been got together, or for their scientific importance, to the series of fossils which he has deposited there. This vast collection has yielded evidence bearing upon the pedigree of the horse of the most striking character. It tends to show that we must look to America, rather than to Europe, for the original seat of the equine series; and that the archaic forms and successive modifications of the horse's ancestry are far better preserved here than in Europe.

"Professor Marsh's kindness has enabled me to put before you a diagram, every figure in which is an actual representation of some specimen which is to be seen at Yale at this present time (Fig. 12).

"The succession of forms which he has brought together carries us from the top to the bottom of the Tertiaries. Firstly, there is the true horse. Next we have the American Pliocene form of the horse (*Pliohippus*); in the conformation of its limbs it presents some very slight deviations from the ordinary horse, and the crowns of the grinding teeth are shorter. Then comes the *Protohippus,* which represents the European *Hipparion,* having one large digit and two small ones on each foot, and the general character of the fore-arm and leg to which I have referred. But it is more valuable than the European *Hipparion* for the reason that it is devoid of some of the peculiarities of that form—peculiarities which tend to show that the European *Hipparion* is rather a member of a collateral branch than a form in the direct line of succession. Next, in the backward order in time, is the *Miohippus,* which corresponds pretty nearly with the *Anchitherium* of Europe. It presents three complete toes—one large median and two smaller lateral ones; and there is a rudiment of that digit which answers to the little finger of the human hand.

"The European record of the pedigree of the horse

stops here; in the American Tertiaries, on the contrary, the series of ancestral equine forms is continued into the Eocene formations. An older Miocene form, termed *Mesohippus,* has three toes in front, with a large splint-like rudiment representing the little finger, and three toes behind. The radius and the ulna, the tibia and the fibula, are distinct, and the short-crowned molar teeth are Anchitheroid in pattern.

"But the most important discovery of all is the *Orohippus,* which comes from the Eocene formation, and is the oldest member of the equine series, as yet known. Here we find four complete toes on the front limb, three toes on the hind limb, a well-developed ulna, a well-developed fibula, and short-crowned grinders of simple pattern.

"Thus, thanks to these important researches, it has become evident that, so far as our present knowledge extends, the history of the horse-type is exactly that which could have been predicted from a knowledge of the principles of evolution. And the knowledge we now possess justifies us completely in the anticipation that when the still lower Eocene deposits, and those which belong to the Cretaceous epoch, have yielded up their remains of ancestral equine animals, we shall find, first, a form with four complete toes and a rudiment of the innermost first digit in front, with probably, a rudiment of the fifth digit in the hind foot;[13] while, in still older forms, the series of digits will be more and more complete, until we come to the five-toed animals, in which, if the doctrine of evolution is well-founded, the whole series must have taken its origin.

"This is what I mean by the demonstrative evidence of evolution. An inductive hypothesis is said to be demonstrated when the facts are shown to be in entire accordance with it. If that is not scientific proof, there are no merely inductive conclusions which can be said to be proved. And the doctrine of evolution, at the present time, rests upon exactly as secure a foundation as the Copernican theory of the motions of the heavenly bodies did at the time of its promulgation. Its logical basis is precisely of the same character—the coincidence of the observed facts with theoretical requirements.

13 Remains of animals corresponding very closely to this description were afterwards discovered.

"The only way of escape, if it be a way of escape, from the conclusions which I have just indicated, is the supposition that all these different equine forms have been created separately at separate epochs of time; and, I repeat, that of such an hypothesis as this there neither is, nor can be, any scientific evidence; and assuredly, so far as I know, there is none which is supported, or pretends to be supported, by evidence or authority of any other kind. I can but think that the time will come when such suggestions as these, such obvious attempts to escape the force of demonstration, will be put upon the same footing as the supposition made by some writers, who are, I believe, not completely extinct at present, that fossils are mere simulacra, are no indications of the former existence of the animals to which they seem to belong; but that they are either sports of Nature, or special creations. . . .

"In fact, the whole evidence is in favor of evolution, and there is none against it. And I say this, although perfectly well aware of the seeming difficulties which have been built up upon what appears to the uninformed to be a solid foundation. I meet constantly with the argument that the doctrine of evolution cannot be well founded, because it requires the lapse of a very vast period of time; while the duration of life upon the earth, thus implied, is inconsistent with the conclusions arrived at by the astronomer and the physicist. I may venture to say that I am familiar with those conclusions, inasmuch as some years ago, when President of the Geological Society of London, I took the liberty of criticising them, and of showing in what respects, as it appeared to me, they lacked complete and thorough demonstration. But, putting that point aside, suppose that, as the astronomers, or some of them, and some physical philosophers, tell us, it is impossible that life could have endured upon the earth for as long a period as is required by the doctrine of evolution—supposing that to be proved—I desire to be informed what is the foundation for the statement that evolution does require so great a time. The biologist knows nothing whatever of the amount of time which may be required for the process of evolution. It is a matter of fact that the equine forms, which I have described to you, occur in the order stated in the Tertiary formations. But I have not the slightest means of guessing whether it took a million of years, or ten millions, or a hundred millions, or a thousand millions of years, to give rise to that series of change.

"A biologist has no means of arriving at any conclusion as to the amount of time which may be needed for a certain quantity of organic change. He takes his time from the geologist. The geologist, considering the rate at which deposits are formed and the rate at which denudation goes on upon the surface of the earth, arrives at more or less justifiable conclusions as to the time required for the deposit of a certain thickness of rocks; and if he tells me that the Tertiary formations required 500,000,000 years for their deposit, I suppose he has good ground for what he says, and I take that as a measure of the duration of the evolution of the horse from the *Orohippus* up to its present condition. And, if he is right, undoubtedly evolution is a very slow process, and requires a great deal of time. But suppose, now, that an astronomer or a physicist—for instance, my friend, Sir William Thomson—tells me that my geological authority is quite wrong; and that he has weighty evidence to show that life could not possibly have existed upon the surface of the earth 500,000,000 years ago, because the earth would have then been too hot to allow of life, my reply is: 'That is not my affair; settle that with the geologist, and when you have come to an agreement among yourselves, I will adopt your conclusion.' We take our time from the geologists and physicists; and it is monstrous that, having taken our time from the physical philosopher's clock, the physical philosopher should turn round upon us, and say we are too fast or too slow. What we desire to know is, is it a fact that evolution took place? As to the amount of time which evolution may have occupied, we are in the hands of the physicists and astronomers, whose business it is to deal with those questions."

INDEX

Made in the USA
Monee, IL
04 November 2026

67384986R00174